THIS IS WHY YOU DREAM

ALSO BY RAHUL JANDIAL

Life on a Knife's Edge

Life Lessons from a Brain Surgeon

THIS IS WHY YOU DREAM

What Your Sleeping Brain Reveals About Your Waking Life

RAHUL JANDIAL, MD, PhD

PENGUIN LIFE

VIKING
An imprint of Penguin Random House LLC
penguinrandomhouse.com

First published in hardcover in Great Britain by Cornerstone Press,
an imprint of Cornerstone, part of Penguin Random House Ltd., London in 2024

First United States edition published by Penguin Life/Viking, 2024

A Penguin Life Book

ISBN 9780593655719 (hardcover)
ISBN 9780593655726 (ebook)

Printed in the United States of America
1st Printing

To Dad, for teaching me how to think.

Contents

Introduction
A Nightly Dose of Wonder

I've spent my life immersed in the brain. As a dual-trained MD, PhD neurosurgeon and neuroscientist, I perform surgery on patients with cancers and other illnesses. I also run a research laboratory. It is impossible to spend so much time treating and studying the brain without being in awe of it. The more I learn, the more I'm captivated, infatuated even.

With billions of neurons and trillions of connections between them, the brain is infinitely complex. But in my own ongoing journey of discovery, one feature of the mind captures my attention like no other: dreaming. For many years, I have searched for answers to the essential questions: Why do we dream? How do wc dream? And perhaps, most importantly, what do dreams mean? In that, I have abundant company.

Dreams have always been a source of mystery. They have captured the attention of humanity's thinkers, from the ancient Egyptians and Aristotle to Charles Dickens and Maya Angelou, from director Christopher Nolan and activist Nelson Mandela to the slain Brooklyn rapper Notorious B.I.G. They inspire invention and art, medicine and psychology, religion and philosophy. They've been seen as omens, messages from the gods and from our subconscious, from the soul and *self*, from angels and demons. They have changed the course of individual lives and the course of the world, spurring marriage proposals and business deals, inspiring song lyrics and scientific breakthroughs, triggering military invasions and mental breakdowns.

Dreams captivate, scare, arouse and inspire us because they are both so real and so surreal. We are simultaneously creators of our dreams and helpless participants in our strange creations. Dreams emerge from us but seem somehow apart from us, home movies we have conjured that do not follow the laws of time or nature, both intimate and out of our control.

As the British poet Lord Byron wrote:

> [. . .] Sleep hath its own world,
> And a wide realm of wild reality,
> And dreams in their development have breath,
> And tears, and tortures, and the touch of Joy;
> They leave a weight upon our waking thoughts,
> They take a weight off our waking toils.[1]

Given the often disjointed and illogical nature of dreams, it may be hard to grasp how the imagined tears, tortures and joys of your dreams reveal much about you. Over time, however, they paint a vivid picture of how we see ourselves and the world. They illuminate our nature, our interests, and our deepest concerns. In this way, we are uniquely our dreams, and our dreams are uniquely us.

Although the creation of dreams may seem mysterious, their source is anything but. The brain reverberates with electricity, waves of current moving across the brain every moment we're alive. Dreams are a product of normal brain electrophysiology, and an extraordinary transformation that occurs in the brain each night when we sleep, following the circadian rhythms—the day-night cycles—that biologically govern all life.

Dreaming should by no means be dismissed just because it

occurs while you're asleep or lacks the logic that guides you during waking hours. Dreams are a different form of thinking. It's their very wildness that gives them the potential to be transformative. Great leaps in art, design, and fashion are built on the type of divergent thinking that comes naturally in dreaming, and it is culture, language, and creativity that has allowed humans to flourish far beyond our physical evolution. Dreaming is at the heart of all this.

Today the word "dream" means many things: ambition, an ideal, a fantasy, and the vivid narratives generated during sleep. Neuroscience is showing that the boundaries between sleep and waking are not so clean after all. Dreams can help you solve a problem; learn a musical instrument, a language, or a dance move; practice a sport; give you clues about your health; and make predictions about the future. Dreams can be spiritually enriching. Forgotten dreams can still shape your mind and influence your day. You can learn to remember dreams, prime their content and even control them during something called lucid dreaming. Most importantly, dreams can offer the greatest gift—that of self-knowledge. By interpreting your dreams, you can make sense of your experience and explore your emotional life in new and profound ways.

Dreams are an elusive form of cognition. Because we experience them alone, closed off from the world, a subjective experience for an audience of one, much about dreams is likely beyond the realm of experimental testing or scientific proof. In this book, I've done my best to capture the current state and breadth of knowledge about dreams and dreaming while noting uncertainties in the research and disagreements among researchers. The book also includes theories I've developed based on the latest research and my own knowledge of the brain. Ultimately, this book is my synthesis of information from disparate disciplines. It is the product of intense effort and even greater humility.

Before we begin, consider for a moment the magic of dreaming. When we dream, we transcend our physical selves. We're no longer aware that we're lying in bed or even lying down at all. Our eyes are closed, but we can see. Our body is still, but we can walk, run, drive a car, fly. We are silent, but we can hold conversations with people we know and love, alive or dead, and people we've never met. We exist in the present, but we can travel back in time, or forward into the future. We are in a single spot but can transport ourselves to places we haven't been to in years, or places that exist only in our imaginations. We are in a world completely of our own making. And it has the potential to be transcendent. Dreams are our nightly dose of wonder.

I.

We Have Evolved to Dream

In the operating theater during a procedure called awake brain surgery, I use a pen-like device to apply tiny amounts of electricity directly onto a patient's brain. Exposed, the undulating surface of the brain is shimmering and opalescent, punctuated with arteries and veins. The patient is conscious and alert yet feels no pain because the brain has no pain receptors. But the electricity has an effect. Every brain is unique and some spots I touch come alive. Touch one spot, and the patient reports a childhood memory. Touch another, and the patient smells lemon. Touch a third, and the patient feels sadness, embarrassment, or even desire.

The purpose of awake brain surgery is to find the precise locations where the flicker of electricity produces nothing. These are the spots where it is safe to cut through the surface tissue to reach the tumor below. When a micro-jolt of electricity produces no response, I know dissecting will not result in any functional damage.

While methodically stimulating the outermost layer—the cerebral cortex—a few millimeters at a time during awake brain surgeries, I have sparked bizarre and profound experiences in patients. Sometimes they are so powerful the patient asks me to stop, and I must pause the surgery. Though the cerebral cortex is less than a fifth of an inch thick, it carries much of what makes us who we are: language, perception, memory, thought. The tiny buzz of electricity can make patients hear sounds, recall traumatic events, experience profound emotion—even dream.

In fact, nightmares can be triggered by electrical stimulation. Remove the probe's current from a certain ridge on the surface of the brain, and the nightmare ends. Return the electricity to the same spot, and the exact nightmare returns. Recurrent nightmares are now understood to be self-sustaining loops of neuronal electrical activity that replay the experience of terror.

In this incontestable way, my craft has answered one of humanity's original questions: Where do dreams come from? I can say with certainty that dreams arise from our brains and specifically our brains' electrical activity.

This basic understanding of the true origin of dreams had long eluded us. For much of human history, dreams were messages from gods or demons or our ancestors, or information gleaned as the soul ventured out in the night. The last place one could imagine dreams coming from was the seemingly inactive flesh in our skulls. The mind during sleep was thought to be dormant, a passive vessel, and dreams were not considered a product of sleep. How could they be? How could our brains, absent any signals from the outside world, be the origin of such nightly brilliance? Something bigger than ourselves, beyond ourselves, must be the source of dreams.

Of course, we now know all consciousness is powered by electricity, including dreaming, and it turns out the dreaming brain is as active as the waking brain. In fact, the electrical intensity and patterns measured during certain stages of sleep look nearly identical to when we're awake. Moreover, the amount of energy certain regions of your brain burn while dreaming can exceed that which they burn when we're awake, particularly in the emotional and visual centers of the brain. While the waking brain might typically adjust metabolic activity up or down by 3 or 4 percent in the emotional, limbic system of the brain, the dreaming brain can boost the limbic system by an astonishing 15 percent. That means dreams can achieve an emotional intensity that is not biologically possible in our waking lives. In fundamental ways, you are never more alive than when you're dreaming.

When we dream, our minds are pulsing with cerebral activity: seeing vividly, feeling deeply, moving freely. Dreams affect us deeply because we experience them as real. The joy we experience in dreams is no different physiologically than waking joy, and so is the terror, frustration, sexual excitement, anger, and fear. There's a reason the physical experiences of our dreams also seem real. Run in our dreams and the motor cortex is activated, the same part of your brain that you'd use if you were really running. Feel a lover's touch in your dream, and the sensory cortex is stimulated, just as it would in your waking hours. Visualize a memory of a place you once lived and you engage the occipital lobes—the area responsible for visual perception.

There are some people who claim they never dream. In reality, virtually everyone dreams, though not everyone remembers. We do not choose to dream. We need to dream. If we're sleep deprived, the first thing we catch up on is dreaming. If we've had enough sleep but are dream deprived, we will immediately start dreaming as soon as we fall asleep. Even when sleep is impossible, vivid dreams can emerge. Among people with Fatal Familial Insomnia, a rare and lethal genetic disease that makes sleep impossible, the need to dream is so strong that dreams escape into the day. Dreaming is essential.

For decades, researchers studying dreaming focused on only one stage of sleep, rapid eye movement or REM sleep. They concluded we spend about two hours a night dreaming, more or less. If you do the math, this adds up to about a twelfth of our lives immersed in dreams, a month out of every year. This would represent an enormous commitment to dreaming. It may also be a gross underestimate. When researchers at sleep labs wake up study participants at different points throughout the night, not just during REM sleep, they are finding that dreaming is possible at any stage of sleep. That means it's quite possible we spend almost a third of our lives dreaming.

There is so much focus these days on the need for sleep to be healthy, but findings like these make me wonder: Maybe it's not the sleep we really need, but the dreams.

What Creates the Dreaming Mind

Dreams are a form of mental activity, but they don't require an external stimulus. They are not triggered by sights, sounds, smells or touch, but arrive automatically, effortlessly. To consider how this is possible, let's take a microscopic look at the brain, starting with the fundamental building block of thought: the neuron.

Neurons form electrical connections in the brain that produce all thought. When we're dreaming, neurons are collectively firing thousands of times per second. Individual neurons are delicate. They are so delicate they need to be protected in a bath of cerebrospinal fluid, which also allows for electrical conduction. This fluid is also rich in nutrients and ions that make neurons a sort of living battery ready to discharge electricity.

In my laboratory and others around the world, we can separate cerebral tissue down to the level of a single cell, or an individual neuron. In a petri dish, a single neuron is alive but inactive. If we were to add a few other neurons, however, the scene changes. The cells will coalesce on their own. They will also do something else, something remarkable. The neurons will start to pass infinitesimal charges of electricity between themselves, and the cellular cluster becomes electric. The surprising thing is that the neurons require no nudging or direction at all. They are receiving no outside stimulus, yet they course with electricity. This amazing interaction is called stimulus-independent electrical activity.

The same thing happens in the brain as a whole, with its 100 billion neurons and their 100 billion supporting cells. They don't sit idling, waiting for the world to excite or provoke them. They have their own waves of electrical activity that flow across the brain, even in the absence of any stimuli. It's called stimulus-independent cognition, and it's why we're able to have thoughts even when we're cut off from the outside world. This is what happens when we dream. Our minds are receiving no external stimuli, yet they are active. But to experience the wild, visual narratives of dreams, three things have to happen.

The first is that the body becomes paralyzed. Our body releases two neurotransmitters, glycine and gamma-aminobutyric acid (GABA), that effectively switch off motor neurons, the specialized cells in the spinal cord that activate muscles. Locking down the body allows it to dream safely. Otherwise, we would be acting out our dreams.

The second thing that has to happen before we can dream is that the brain's Executive Network must turn off. The Executive Network is composed of structures on both sides of our brain that co-activate and are responsible for logic, order, and reality testing. With the Executive Network switched off, we can ignore the normal rules of time, space, and reason. Because we have temporarily cast aside reason and logic, we can also accept our dreams' improbable plots unquestioningly. This gives dreams both their power and their unique character.

The third thing that happens when we dream is our attention turns inward. When this occurs, we activate widely dispersed and disparate parts of the brain, collectively called the Default Mode Network (DMN). The name Default Mode Network is misleading, as it is anything but a passive default. For that reason, in this book I'll refer to these associated regions of the brain as the Imagination Network, an alternate name used already by some in the scientific community, because of the connection between the brain network and imaginative thinking.

When we're awake but our mind is not engaged with an activity or task, it is not blank, like a computer with a blinking cursor awaiting a command. Instead, the brain naturally pivots from its Executive Network to its Imagination Network, from directing our attention outward to focusing inward. With the Imagination Network engaged, the mind wanders freely, a meandering path that often leads to unexpected insights. When the outside world doesn't warrant our attention, the regions of the brain that make up the Imagination Network reign.

As we go through our day-to-day life, the Executive Network and Imagination Network essentially take turns being dominant. Right now, as you read these words, the Executive Network is engaged. But the Imagination Network isn't derelict. It wants to jump in, waiting for a break in the tasks that preoccupy the Executive Network. When that happens, our attention turns inward and the Imagination Network comes to life. When the Imagination Network is active and assumes the top position in our cognitive hierarchy, it looks for loose associations in our memory and out-of-the-box connections tied together by the thinnest thread and visualizes what-if simulations. These can be so fanciful or far-fetched our rational brain might dismiss them out of hand when the Executive Network is in charge. Thanks to the Imagination Network, our dreaming brain is unbound and promiscuous in a way our waking brain is not and could never be.

The Imagination Network is central to the experience of dreaming. It allows us to "see" without receiving visual information from the outside world. In fact, shine a bright light into a dreamer's eyes, and they are blind to it. When we dream, it's like a movie playing in a dark theater. This is no doubt why the ancient Greeks described the experience of dreaming as "seeing" a dream, rather than "having" a dream.

When the Imagination Network is engaged, spontaneous thought emerges. Just as clusters of neurons in a petri dish come to life with electrical activity without any external stimulus, the dreaming brain is alive with electrical activity, even though it is largely shut off from the world around us. That's why the Imagination Network has been dubbed the brain's dark energy. It creates something from nothing, fashioning stories from thin air.

Edward F. Pace-Schott, a professor of psychiatry at Harvard Medical School, described the Imagination Network as a veritable storytelling instinct because it spins memories, characters, knowledge, and emotion into coherent narratives.[1] These free-flowing stories are created from nothing and yet are imbued with meaning. When faced with a gap in reality, the human brain creates a coherent narrative to fill the gap. Patients with certain types of partial amnesia will do the same thing. Rather than saying they don't remember if they're asked a question that lands in a gap in their memory, they will casually make something up. People with Alzheimer's disease will sometimes do this as well.

Powered by the Imagination Network, dream narratives flow effortlessly. Though we create our dreams, we rarely have the experience of being able to will the movements that occur in them. In that sense, we are more leading actors than directors. But this should not be confused with being in a dissociative state, floating above and apart from the dream narrative. It's more like being in the driver's seat of a car we don't control. We are still the protagonist in our dreams and fully inhabit the dream experience. We just don't consciously steer where dreams are heading.

When we dream, we are fully embodied in the dream and separate from other characters in the dreamscape. The dream *self* has a physical presence. That doesn't mean our dream body is the same as the one we inhabit when we're awake. Our dream body may be younger, older—even a different gender. We also have a sense of being apart and unique from others in the dream, even though all the characters in the dream are the product of our imagination.

In our dreams we weave a narrative as we move through disparate memories, and our dream *selves* act and react. It's quite a production. We may respond in ways that are different from our waking *self.* We may be stronger or weaker, more assertive or more passive. In this sense, we could consider ourselves as having a waking *self* and a dream *self*, or *selves*.

But how unique is the dreaming brain, really? After all, we are also at the center of our daydreams. Like dreams, we can picture imagined scenarios when we daydream, and our minds can flit from topic to topic, making leaps in time and place. Daydreams, however, are different. Daydreaming is directed thought: Wouldn't it be nice to vacation in Hawai'i? What would happen if I quit my job?

How about psychedelic drugs then? They also produce what

is often described as a dreamy experience, but this, too, is different from dreaming. With psychedelic drugs, the Imagination Network is actually less active, unlike its supercharged state in the dreaming brain. And unlike dreams, where the dreamer is the central character in the drama, the experience on psychedelics is disembodied and dissociative.

If there is any waking state that partially overlaps with dreaming, it is mind wandering. When our mind wanders, thoughts arise one after another without being oriented toward any particular task or goal. In fact, we are not directing our thoughts toward anything at all. Even though neither mind wandering nor dreaming are goal-directed, there are differences. Mind wandering still remains bound by most of the strictures of the Executive Network. The wandering mind is somewhat liberated, but not to the extent of the dreaming mind. The unfettered nature of dreaming is able to take us places when we're asleep that are impossible in our waking life.

Even Dreams Have Rules

As wild and untamed as dreams are, with implausible situations and irrational jumps in time and place, there are limits—even dreams have rules. Though the Imagination Network unleashes the dreaming mind, dreams are not infinitely wild, and they are anything but random. When you zoom out from one dreamer to 10,000, and a single dream to thousands upon thousands of dream reports and descriptions of dreams dating back to antiquity, contours emerge. For instance, despite massive changes in the way we live, the content of dreams has changed little through the ages, from millennium to millennium and generation to generation.

Many common dreams today are no different than those dreamed in Egypt in the time of the pharaohs, or Rome in the time of Caesar. Sleep disorders recorded in China more than 1800 years ago include dream flying, dream falling, and night terrors. Sound familiar?

Questionnaires given to Japanese and American college students in the 1950s showed just how universal dreams really are. Students in these two countries were asked "Have you ever dreamed of . . . ?" with a list of possible dreams, including everything from swimming and being nude to being buried alive. The similarity in the answers among students half a world apart was astounding.

The top five dreams experienced by the Japanese students were:

1. Being attacked or pursued
2. Falling
3. Trying again and again to do something
4. School, teachers, studying
5. Being frozen with fright

Among Americans, the top five dreams were:

1. Falling
2. Being attacked or pursued
3. Trying again and again to do something
4. School, teachers, studying
5. Sexual experiences (sexual experiences were the sixth most commonly experienced dream among the Japanese students surveyed)

Fifty years later, a similar survey was given to students in China and Germany. They also came up with remarkably similar answers.

The top five dream reports among the Chinese students were:

1. Schools, teachers, studying
2. Being chased or pursued
3. Falling
4. Arriving too late, e.g., missing a train
5. Failing an examination

Among the German students, they were:

1. Schools, teachers, studying
2. Being chased or pursued
3. Sexual experiences
4. Falling
5. Arriving too late, e.g., missing a train

How could dream surveys, done half a century apart in four separate countries, produce such similar results? Perhaps it's related to daily experience. After all, the United States, Japan, Germany, and China are all modern, industrial societies. Maybe the lives these students lived were similar enough to produce similar dreams. Would the dreams of people living in indigenous cultures be different?

Anthropologists in the 1960s and 70s decided to find out. They gathered dream reports from indigenous peoples such as the Yir Yoront in Australia, Zapotec in Mexico, and Mehinaku in Brazil. They compared the characteristics of their dreams to those of American dreams, focusing on themes like aggression, sexuality, and passivity. Despite the enormous differences in the lived experience of the traditional cultures and the Americans, the dreamscapes were much more closely aligned than the cultures that produced them.

For example, dream reports from traditional societies and

the U.S. both found men more likely to dream about other men, while women dreamed of men and women equally. In both cultures, men and women were more likely to be the victim of aggression, rather than the aggressors, while fewer than 10 percent of dreams were sexual, another match.

Dreams are remarkably similar all over the world, regardless of what language we speak; whether we live in a city or a rural area, a developed country or a developing one; regardless of wealth or standing in the world. Given this continuity of dreams across time and place, it seems reasonable to conclude that the characteristics and contents of dreams are baked into our DNA, a function of our neurobiology and evolution, largely immune from differences in culture, geography, and language. In the pages that follow, we need to keep this central fact about dreaming in mind: Dreams exist within the framework of their neurobiological origins. As such, they are not truly limitless. As magical as dreaming seems to be, dreams hew to certain boundaries.

Dreams follow rules in other ways. Math doesn't play a role in your dreams, for instance, and it's rare to use other cognitive processes like reading, writing, and using a computer while dreaming. Without the logic of the Executive Network, these are difficult, if not impossible.

You also likely never dream of a mobile phone riding a horse, for example, and it's extremely rare for objects to turn into people in our dreams, or vice versa. In Shakespeare's *A Midsummer Night's Dream,* characters transform into animals, but humans are rarely transformed into animals in dream reports. When objects turn into other objects, they are likely to turn into something similar. A car turns into a bike. A city bus turns into a school bus. A house turns into a castle, or a house in one place turns into a house somewhere else. The jumps in dreams follow the semantic maps in our memory.

Semantic maps are how we organize the people, objects, and places that populate our world. You can think of the semantic maps like clusters of grapes. One cluster might be modes of transportation. Another might be types of dwelling. As your dreaming mind leaps from association to association, it tends to stay in the same semantic cluster. One mode of transportation turns into another. One type of dwelling turns into another. As far as we can tell, this is the way dreams have been for as long as humans have been recording them.

The Social and Emotional Power of Dreams

I wonder if dream narratives have maintained a remarkable consistency over the span of human history because they tend to focus on emotion and interpersonal relationships, both real and imagined. The dreaming mind plays out all sorts of what-if scenarios without any sort of judgment. This is why you can dream yourself a different gender, a different sexual orientation, and place yourself in sexual or interpersonal situations that would be unlikely—even distasteful—in your waking life. We do this largely through the lens of emotions: How would it feel if I did this?

The emotional and social focus of dreams is likely also why dreams don't seem to be much affected by technologies that have transformed life since the 1950s. Television, computers, the internet, and smartphones are rarely found in dream reports. Even our current addiction to social media does not appear to have invaded the dreamscape, based on the limited but growing investigation into how our digital lives populate our dream life.

What the imaginative world of dreams gives us first and foremost are social experiments. We are social creatures. Dreams provide thought experiments that probe the relationships in our lives, often implausible and other times profoundly moving, building our social intelligence in the process. This central feature of dreaming relies on the most recent and most prominent evolutionary advance in the human brain and the Imagination Network, the medial prefrontal cortex (mPFC).

The mPFC sits at the midline of your brain and is a tuft of neurons in part of each frontal lobe, left and right, behind your forehead above the bridge of your nose. Prefrontal means the very front of the frontal lobes, which places them directly behind our forehead. The prefrontal cortex is what has pushed the human forehead forward. This is an area where the newest neurons are being cultivated, revealing evolutionary pressures making us more social, more human.

In our waking life, the mPFC plays a role in our ability to consider both our own point of view and the point of view of others. This is an extraordinary ability. Even as the human brain has gotten smaller in the last 3,000 to 5,000 years, our social intelligence as a species has increased. For this, we have the mPFC to thank. Damage to the mPFC results in lack of empathy, poor social decision-making, and a failure to follow social conventions. It also makes it difficult to change your initial judgment of someone, even after you receive new information.

As we dream, the mPFC is liberated when the Executive Network steps back and the Imagination Network takes center stage. When we attribute thoughts, feelings, and intentions not only to our dream selves, but to the other characters we invent in our dreams, it is as a result of the mPFC. This ability to put yourself in someone else's shoes, particularly in relation to you, is given the shorthand Theory of Mind.

Theory of Mind allows us to consider our beliefs, desires, and emotions and infer those of the people we are interacting with. Ascribing mental states to ourselves and others begins in childhood and is considered vital to our ability to successfully function in a tribe, community, or society. People with such conditions as autism, schizophrenia, and social anxiety disorder have trouble with this, making interactions difficult. Theory of Mind helps us understand why someone acts the way they do, and how they might act in the future. As we dream, Theory of Mind allows us to think about how we would feel in certain imagined situations and how others would feel about us in those same scenarios. This is important because it improves our ability to interact in groups, to solve problems collectively, and to work with a shared purpose. Theory of Mind is in full force in the dreaming mind, letting us play out complex social scenarios, imaginative thought experiments that can inform our waking lives.

As we run through these thought experiments in our dreams, we also have access to a super-charged limbic system. The limbic system is responsible for emotion, memories, and arousal. You will recall that the limbic system during dreaming can be activated to levels that are impossible when we're awake. This hyper-activated emotional state can improve our social intelligence and insight. If you're wondering how emotion can be so crucial to our social skills, keep in mind that when the limbic system is injured and the rational executive part of the brain does not have access to it, our thinking becomes paralyzed and unable to make sense of the social world or even make straightforward decisions. Damage to the limbic system can impede the ability to empathize, understand social cues, and interact appropriately with others. Though we don't typically think of them that way, emotions are essential to optimal judgment in social situations. I believe this capacity has driven our collective evolution forward.

The Dreaming You vs. the Waking You

Most of us have a clear sense of who we are. Beyond our physical appearance, we have memories of what we've done in the past and ideas about where we'd like to see ourselves in the future. We have beliefs and morals, likes and dislikes. All of this paints a detailed self-portrait. But what about the protagonist in your dreams? Is your dream-*self* different from who you are when you're awake?

In the middle of the twentieth century, American researchers Calvin Hall and Robert Van de Castle developed a system for breaking down dreams into their component parts.[2] This coding technique scored how many characters were in a dream. Were they individuals, groups, or animals? Were they male or female? How aggressive was the dream? Was the dreamer the aggressor or the victim?

They found you will almost always be the main character in your dream, the plot of your dreams will typically contain about five characters, and dream narratives are more likely to skew toward misfortune, rather than good fortune, and toward aggression, rather than kindness. Using this scoring system, Hall, Van de Castle and others also showed most dreams are not bizarre, but the ordinary stuff of everyday life.

The notion that dreams are a continuation of our waking lives is called the continuity hypothesis of dreaming. The continuity hypothesis doesn't say our dreams perfectly mirror our waking lives, but it does say they reflect our personality, values, and drives, and that our dreams carry over from emotional preoccupations and concerns or needs when we're awake. According to proponents of this theory, perhaps as many as

70 percent of our dreams amount to "embodied simulations" of concerns and conceptions that are personal to us.[3]

Anyone who has incorporated their boss into a dream after a tough day at work or a beloved relative not long after their death knows elements of our lives make it into our dreams. A study comparing mothers who worked outside the home with stay-at-home mothers found that mothers with jobs experienced more unpleasant emotions, more male characters, and fewer residential settings in their dreams than women who stayed home.

Yet we all know dreams are often nothing like our waking lives. It seems to me there is at least as much discontinuity as continuity. Much of what is represented from our waking life in dreams is distorted or taken out of context. It is often a strange cocktail of the real and unreal.

Researchers have tested how much of our day-to-day reality is incorporated into dreams by drastically altering the lives of their research subjects. Studies have used colored goggles, immersive video games, and other techniques to see how our waking reality seeps into dreams. As you might imagine, it's rarely a faithful representation. People who wore red-tinted goggles all day sometimes dreamed in red, or sometimes only part of their dreams were "goggle colored."[4] In another experiment, participants wore "inversion goggles" that turned the world upside down.[5] They didn't wind up dreaming of this inverted reality, but their dreams would incorporate some inverted things. Elements from video games appear in dreams, but the dreams are rarely a replay of the game. That would be too prosaic for the dreaming brain.

Over time the patterns of our dream narratives are unique to each of us, but they shouldn't be expected to faithfully replicate our daily lives. Hall and a colleague analyzed 649 dreams from an American who gave herself the pseudonym Dorothea. Dorothea started recording her dreams in a diary in 1912, when she was twenty-five years old, and continued until a few days before her death in 1965, at age seventy-eight. In her dream reports over five decades, a handful of themes dominate, appearing in an astonishing three-quarters of all her dreams:

- Food and eating
- Losing an object
- Being in a small or disorderly room or having her room invaded by others
- Being in a dream with her mother
- Going to the toilet
- Being late

This pattern of dreaming showed remarkable consistency decade after decade. You might be able to read 100 or 200 of Dorothea's dream reports and know they belonged to her. But these dreams do not actually give us any clues about her life. You would never know from these dreams that she was the second of eight children, or born in China to Chinese missionaries, or returned to the United States when she was thirteen, or earned a PhD in psychology at thirty-eight, that she never married or had children and taught until she retired. The best you could hope to learn from Dorothea's dreams was inferring something about her values, her concerns, and her preoccupations.

Hall himself had trouble gauging individual personalities and characters from his patients' dreams. Studying the dreams of seventeen men on the 1963 American Mount Everest Expedition, he decided two of them would be the most popular, the most psychologically mature, and the best leaders. It turned out he was completely wrong. They were the least liked, the most immature, and were considered terrible at leading or building morale. Hall wrote that he was sobered by "the enormity of misjudgments" he'd made trying to determine waking behavior from the content of the climbers' dreams. Hall's misjudgment showed the limits of dreams to reflect our waking reality. Dreams appear to be a distorted mirror, at best.

How Dreams Develop in Childhood

Though my three sons are now in college, I remember watching their development as infants and toddlers. The first real smile, the first word, the first step, the first day of preschool. Like most parents, I was both excited and relieved as they reached each of these developmental milestones. As a young child grows and experiences the world, the brain has other equally significant neurological accomplishments that occur out of view of even the most attentive parent. Though these milestones happen "off stage," they are no less significant, especially when it comes to dreaming.

The ability to dream is a significant cognitive achievement that takes time to develop. In fact, we're walking and talking before we're dreaming. We develop the capacity to dream in step with the development of visual-spatial skills around the age of four, about the same time we're learning how to hop, balance on one foot, and catch a ball.

We know about the dreams of children over time because of longitudinal studies that have tracked the arrival and evolution of their dreams. In some cases, children and their families have even participated in dream reporting and evaluation over the course of decades well into adolescence and adulthood. As a result of this intensive research, we know the dreams of children and their waking imagination grow in tandem.

The earliest dream reports by children barely qualify as dreams. Children between the ages of three and five awakened during a stage of sleep when dreams are abundant in adults usually do not report dreaming at all. And if they are dreaming, the dream doesn't involve motion or movement. They are more like still photos than video. There is very little movement, very little social interaction, and the dreamer is usually not participating in the dream.

Among preschoolers, aggression, misfortune, and negative emotions are rare. The two main characteristics of dreams at this stage are animal characters and references to the state of the body, for example hunger or fatigue. A dream centered on the bodily state might be one of sleeping at the kitchen table, while an animal dream might consist of a bird chirping. Interestingly, animals in the dreams of young children typically don't involve their own pets but animals from fairy tales, cartoons, and stories. One hypothesis for this is that the animal characters serve as a stand-in, a sort of avatar for the child before their sense of *self* is fully developed.

From ages five to eight, children start reporting narrative dreams, though without chronology or sequence. Children initially think dreams are shared fantasies, but eventually they come to realize their dreams are not a shared experience but something private. This happens in parallel with the Imagination Network coming online around this age. The brain structures of the Imagination Network take time to connect with each other and function in concert for its specific behavior or purpose.

But it's only around age seven or eight that children become active participants in their dreams. At the same time, their dream reports begin to show a sequence of events, with one event leading to the next. This is the same time in a child's life when an awareness of an autobiographical *self* also emerges in both dreaming and waking life. The autobiographical *self* is a sense of who we are, both in ourselves and in relation to others. Given the confluence of these developmental events, it seems likely they are related, perhaps influencing or promoting one another.

What is it that finally gives children the ability to dream? If you think about it, most children are already going to school and learning to read or do simple math, but they aren't yet dreaming, at least not in the way we think of dreams as a series of scenes. This puzzled researchers who wondered whether young children were having dreams all along but simply didn't have the verbal skills to describe them. But this explanation doesn't make sense, given that children have the ability to talk about people, events, and things before they report dreaming about them.

The reality is, the arrival of dreams the way most of us think of them occurs with the development of visual-spatial skills, not language and memory skills. Dreaming requires a lot of us. Not only do we need to visualize the world, we need to manufacture situations. Dreams are like other high-level cognitive processes that come with age and maturation. The key to the ability to dream is how well our minds can visually recreate reality. There's actually a test you can give a child to determine whether they are capable of dreaming called the Block Design test. In this test, children must look at models or pictures of red and white patterns and then recreate those patterns in blocks. If they can match the pattern, they can probably dream.

Both visual-spatial skills and dreams depend on the parietal lobes, which help with spatial orientation and aren't fully developed until around age seven. More importantly, dreams rely on complex associations between brain regions, the association cortices, that also take time to develop and make meaning out of what the occipital lobe may see and the parietal lobe may feel—working together for an immersive visual and emotional experience.

Soon after the arrival of dreaming, a strikingly universal event in pediatric development occurs: nightmares. We'll look at nightmares more closely in the next chapter, but children experience many more nightmares than adults. Their dreamscapes are populated with monsters and supernatural beings no matter how benign their rearing. Nightmares then fade for almost all of us as we move from childhood to adulthood.

Consider this: We now know dreaming corresponds with our development of a sense of *self*—the essential capacity that allows for an autobiographical memory and identity. No dreams serve to reinforce the sense of *self* more than nightmares. In a nightmare, the *self* is typically under attack or facing some other kind of existential threat. A nightmare is essentially a battle of *self* versus other. This is a powerful way to instill the notion in a child that they are separate beings, with their own will and their own place in the world.

The Evolutionary Benefits of Dreaming

How do we know dreams aren't random? Couldn't they be a series of images, memories, characters, and actions pulled like cards from a deck? Dreams could be the unimportant byproduct of something beneficial that happens during sleep, the noise of an engine but not its pistons and gears.

There are a couple of straightforward reasons we know dreams aren't random. One is that many of us have recurring dreams. If dreams were random, the chance of having the same dream twice would be exceedingly low. The chance of having the same dream more than twice would be impossible. Secondly, some of us can get up in the middle of the night, return to bed, and resume the same dream we were having. This would also be impossible if dreams were truly random.

I believe we have evolved to dream. Here's why. Whenever possible, evolution holds onto traits that are advantageous. Evolution would never perpetuate traits that did not give us a clear advantage, especially if they demanded a lot of energy or exposed us to predation. Dreaming does both. It demands a lot of energy and leaves us vulnerable while we are dreaming.

So why do we dream? Why go through these nocturnal exertions, these bizarre, mental narratives conjured only for ourselves—of falling, of having our teeth fall out, of cheating on a partner? What possible biological or behavioral advantage do we get by spending years, maybe decades, of our lives dreaming?

These questions have provoked a lot of theories. We all dream at some time of being chased, so one theory is that dreams exist as a kind of threat rehearsal, a way to practice recognizing and responding to threats in a safe way. In this theory, dreams are like a virtual simulation where we can test different responses and imagine the consequences. Could it be that we are better at managing real-world threats based on our dream life experiences?

In perhaps a modern version of threat rehearsal, Isabelle Arnulf, a professor of neurology at the Sorbonne University in Paris, asked students about their dreams before their medical school entrance exam.[6] Dreams about the exam were common and more than three-quarters were nightmares. The themes of these unpleasant dreams are ones you could no doubt predict: "I woke up peacefully at 10 a.m. Suddenly, I panicked and realized it was all over and I failed the exam." Other students dreamed of eyeglasses shattering before the exam, receiving an exam with pages missing, having no paper for writing during the exam, missing the exam because the train went in the wrong direction, and so on.

Interestingly, students who dreamed about the exam frequently did about 20 percent better than those who never dreamed about it. More sleep did not predict better results, nor did higher pre-test anxiety forecast a lower score. Arnulf concluded that negative anticipation of a stressful event and the simulation of the test during dreaming may have given the test takers a cognitive benefit. The dream reports, she concluded, served as a kind of checklist for all possible situations, ranging from those that were likely, such as forgetting one's documents, to those that were improbable or impossible, such as taking a plane to the exam.

However, if threat simulation was the only reason we dream, then all our dreams would involve imagined threats. We know that's not true. Dream plots are varied, and we experience many emotions besides fear when we dream. There must be other evolutionary benefits to dreaming.

Another theory suggests dreams have therapeutic value, serving as a kind of nocturnal therapist, helping us digest and metabolize anxiety-provoking emotions. Many of us have dreamed of being late or showing up underdressed or naked in public. These dreams can actually help us avoid embarrassment in our waking life. Recent research at the University of California, Berkeley, shows fear responses to emotional waking experiences are blunted the morning following these kinds of dreams.[7]

Evidence of the therapeutic value of dreams can also be found in the dreams of divorcing couples. Rosalind Cartwright at Rush University Medical Center's Graduate College of

Neuroscience, in Chicago, found dreams could by themselves be accurate predictors of who would—and who would not—recover from post-divorce depression.[8] Those who recovered tended to have dreams that were more dramatic, with complex plots that mixed old and new memories. Cartwright concluded the recently divorced test subjects were working out their negative feelings about their former spouse in their dreams. This, she said, had the effect of defusing the emotion and preparing the dreamer to wake up ready to see things more positively and to make a fresh start. The degree to which divorcing couples dreamed about each other was correlated with how well they moved on.

Dreaming may also serve as a means of testing different interpersonal scenarios. When it comes to visualizing all sorts of social situations, nothing matches dreaming. Dreams are capable of delivering an incredible range of plots, both realistic and implausible, and in each of them, we imagine how they play out. We're so good at this, social scenario building has justifiably been dubbed our "superpower" by Mark Flinn, a biomedical anthropologist at Baylor University in Texas.[9] How well we interact with others is vital from an evolutionary perspective. It helps us get along in a group and find a mate.

Yet another theory about the evolutionary benefits of dreaming points to keeping the brain tuned and ready even during sleep. In trying to create machines that behave like the mind, computer scientists encounter challenges that give us clues into other benefits dreams could provide.

Neural networks are functionally associated neurons in the brain. For example, a neural network could be the type of visual processing required to determine whether someone we see is familiar to us. Facial recognition software is an artificial version of this. One theory proposes that dreaming has evolutionary benefits because the accompanying bursts of mental activity keep neural networks finely tuned, as a sort of pilot flame for the brain. That way, if we are awakened, the brain can quickly become alert and engaged.

Machine learning and the bizarre nature of dreams have inspired still one more theory of the evolutionary benefits of dreaming. Dreams are often surreal, with outlandish or improbable situations, the type you wouldn't see in a typical day and may never see in your lifetime. With this in mind, American neuroscientist Erik Hoel proposed something he calls the overfitted brain hypothesis.[10] He suggests dreams exist to help generalize what we've learned in our waking life.

When a machine learns complex tasks, it is trained to develop general rules from a set of specific circumstances. If the specific circumstances used are too similar, "overfitting" occurs, and the rules the machine adopts become too closely aligned to the limited information it has received about the world. As a result, the machine becomes what in humans would be considered narrow minded. The machine's thinking is too zoomed in, too rigid and formulaic in its analysis. In other words, the machine will fail when it receives data that is "outside the box." To prevent this from happening, computer scientists inject "noise" into the information used to teach the machine, deliberately corrupting the data and making the information more random.

Like the data sets a machine receives for machine learning, our day-to-day lived experience can often give us limited information about our world, creating patterns of thought that are constrained or boxed in. Habituating to a routine is efficient but also limits how adaptable we are to unexpected circumstances. Dreams, with their fantastic, often surreal quality, are much like the noise injected into the machine's data. This nightly reshuffling of our memories and patterns could be relying on something called stochastic resonance, a scientific term that describes adding random noise to data to make important signals more detectable, rather than less so. This could lead to more flexible and creative thinking.

It's not just the mind and outlandish dream narratives that support this theory, but actual neurophysiological changes that occur during dreaming. The brain injects "noise" into our dreams by lowering adrenaline levels. We're familiar with adrenaline because that's the neurochemical that kicks our fight-or-flight response into high gear and makes us acutely vigilant. More adrenaline puts us in a hyper-alert, hyper-focused state. When that happens, we are best at detecting even the faintest signal from the noise. This had enormous benefits when humans were avoiding predators in the wild. A boost of adrenaline might help us detect a faint rustle in the tall grass alerting us to the presence of a threat just out of sight.

When we're dreaming, adrenaline is turned down, and our discernment between signal and noise loosens. As a result, the brain has dampened reality testing. This would be a massive vulnerability if we were facing danger, but it gives dreaming the power to think creatively and divergently. I'll get into the anatomical and biological basis of divergent thinking more in Chapter 4, on Dreaming and Creativity, but when I write about divergent thinking, I mean what is often called outside-the-box thinking. This is the kind of thinking that looks at a problem in a completely novel way, or from an original perspective, and that can be very difficult when we are knuckling down to solve a problem during our waking hours.

The lack of adrenaline in our brains during dreams allows for the suspension of disbelief required for this sort of adventurous dreaming. This is part of the Executive Network being turned off, step two of dreaming. This makes sense, as it is a chemical synergy of sorts. The Executive Network and adrenaline in the brain both serve similar functions: vigilance and outward focus. At the same time, the adrenaline in our bodies is unchanged and will cause us to experience dreams as though they are real. When we're dreaming of fleeing a predator, for example, the adrenaline in our bodies will make our hearts race as though we're really running for our lives.

This type of imaginative and unbound thinking during dreaming could pay dividends by finding adaptive solutions to existential threats. When we talk about evolution being survival of the fittest, I believe fittest means the most adaptive. Dreams' bizarre narratives help us do just that: navigate the world with all its complexity and give us the best chance of dealing with the widest set of challenges we may face. Dreams can simulate black swan events we would never predict from our daily routines, but which the species might need to react to in order to survive—plagues, earthquakes, tsunamis, war, drought.

At the end of the day, even as the research has blossomed, no theory has emerged as to the single reason humans have retained the need to dream. In fact, the evidence suggests all these theories are valid to some degree, intertwined and interdependent. We shouldn't expect there to be a single reason for dreaming, just as there isn't a single reason for waking thought. As humans have evolved, as the brain added new and more sophisticated layers to our cellular architecture, why couldn't we add to the armamentarium of dreams? Why can't dreams both help us with our emotions and simulate worst-case scenarios? Why can't they serve as threat simulation and keep the neural network finely tuned?

These theories explain all the ways in which dreams help us as a species to adapt and survive, but I believe dreams also help us become who we are. One particular dream appears to play an outsized role in the cultivation of our sense of narrative identity and sense of *self*, allowing the unique person to emerge. It's one we've all had: the nightmare.

2.

We Need Nightmares

Julia was living a tranquil life by day. She taught yoga and spent time gardening and hiking. Yet for years, she had horrific and violent dreams that seemed to come from nowhere, like seeing her parents beheaded, or stabbing someone with a knife. As she recounted in the podcast *Science Vs*, she'd wake up shaking, the disturbing details difficult to forget.[1] When the day began and some of the emotions of her grisly dreams faded, she couldn't help but reflect on the macabre scenes her brain could conjure at night. Increasingly, the nightmare residue lingered into the following day.

Julia was living a perplexing double life. Her days were filled with positive emotions built around habits of wellbeing, but her nights were filled with imagined violence. She was deeply troubled that such violent thoughts lurked within her. She couldn't understand why she was having these nightmares—or what she could do to stop them.

How could Julia's dream life be so different from her waking life, and how could it turn so macabre? Where did these violent nightmares come from?

Indigenous cultures attributed nightmares to external forces: evil spirits, demons, or other malevolent beings. Some cultures don't even have a term for nightmares and instead consider them windows to the edges of consciousness. The truth is that nightmares, like dreaming, are the product of neurobiology. Ultimately, nightmares' dark visions are our own.

To many of us, nightmares seem like an unwanted side effect of sleep. After all, they fill us with heart-pounding terror and wake us up. You may dread them. They may haunt you. Yet, they are necessary—beneficial even—in ways you may never have imagined.

To understand nightmares, it's helpful to think of them based on the age of a person when they occur, their origin, and the role they may play. Of course, no feature of the mind can be neatly divided, but these distinctions are helpful to begin our exploration. The type of nightmares this chapter focuses on are those that occur for all of us in childhood and persist for some into adulthood. Their universal appearance may serve a purpose in children to cultivate their identity and sense of *self*. They are filled with terror but rarely disrupt the child's life.

Another type of nightmare, usually experienced by adults, not only terrorizes our dreams but affects waking life and serves as a psychological thermometer of sorts. They can be brought on by stress or anxiety and trauma itself. If they are severe enough or chronic enough, they can qualify as something called nightmare disorder. We will take a look at trauma-induced nightmares in Chapter 5, Dreaming and Health.

But first, what distinguishes a nightmare from other dreams?

More than a Bad Dream

Nightmares shouldn't be confused with a bad or unpleasant dream. A bad dream is merely one we would rate as emotionally negative: You miss the bus, or have to interact with an unpleasant colleague. Nightmares, on the other hand, are characterized by long, vivid, frightening dreams that always wake us up.

The nightmare plot usually involves a threat to our survival, physical integrity, security, or self-esteem, and their emotional atmosphere is one of dread. They can also produce intense feelings of fear, anger, sadness, confusion, and even disgust. By definition, nightmares force us not only to wake but to vividly recall the frightening events.

The content of nightmares differs greatly from that of the other main categories of dreams, the pleasurable dream and the dream in pursuit of a goal. These dreams tend to be more metaphorical than literal, while nightmares are often more literal than metaphorical. In a nightmare, we are typically menaced in some real way. Our dream *self* comes under attack.

Nightmares are different in another way. In other dreams, we are often able to infer the motives and emotions of the other characters; in nightmares, we can lose this mind-reading ability. Faced with a realistic threat from an unreadable foe, your sense of *self* is heightened. In nightmares, it is you versus an "other."

A popular and persistent myth says it's not possible to die in a dream, or that if you do, you'll die in real life. The source of this misguided lore is unclear, but it has persisted across generations. The truth is you can die in a dream, but you almost always wake up before that happens.

Even if the content of dreams can't kill you, the physiological stress of our most emotional dreams can. Approximately every ninety minutes, we go through the full cycle of sleep: light sleep, deep sleep, and finally rapid eye movement (REM) sleep, where we experience our most vivid and emotional dreams. With each sleep cycle through the night, the period of

REM sleep gets longer and the dreams become more emotionally intense. It should come as no surprise then that the final period of REM sleep before waking is associated with an increased risk of cardiac arrest.

During a nightmare, the amygdala, the part of the brain that processes emotional experiences, throbs with activation. Our breathing can become rapid and irregular, we may start sweating, and our heart rates can spike. One person's heart rate during a nightmare was recorded skyrocketing from sixty-four beats a minute to 152 beats in a mere thirty seconds. Yet most nightmares leave no lasting mark on our bodies, even when their content is burned into our psyche.

No matter how nightmares disturb you, shake you, or change you, they remain to a large degree a cipher. The source of their disturbing power is hard to pin down and quantify. They are a subjective, private, visual and emotional roller-coaster experienced during sleep, and evaluated by our subjective consciousness when we're awake.

Nightmares are universal, and, as far as we can tell, have always been part of the human condition. They are not a glitch or aberration, randomly affecting some people and not others. Everyone has nightmares. They aren't limited by life experience, or diet, or age, or personal habits. The gentlest of childhoods is no barrier against nightmares.

Nightmare themes are also not random. They aren't a sporadic firing of neurons with ominous organ music playing in the background. The plots of nightmares are predictable. The five most common themes across the world and over time are: failure and helplessness, physical aggression, accidents, being chased, and health-related concerns or death. Nightmares are often our earliest remembered dream, and each of us could probably name a nightmare that has recurred periodically throughout our lives, startling us awake and shaking us to the core.

Children Have the Most Nightmares

Have you ever wondered why you needed to have a nightmare in the first place? What possible benefit could they offer? I believe nightmares may help us in a number of ways, not only as individuals but as a species. The most important benefit comes early in life and might surprise you.

Nightmares unfold throughout our lives in an intriguingly predictable pattern. For starters, children experience nightmares an estimated five times more often than adults. Childhood nightmares often involve falling, being chased, and an evil presence. In dream reports throughout the world and across all cultures, children dream of monsters, demons, and supernatural beings. How can this be? How can children reared with love, who are nurtured and protected, still conjure up monsters?

Proving beyond a shadow of a doubt how and why this feature of childhood came to be may never be possible, but when you consider the patterns and themes of nightmares, it's tempting to speculate.

Let's first consider the soil in which these terrifying dreams proliferate. Childhood nightmares arrive at a time of explosive cognitive growth. Language and social skills are flourishing. As young children interact at home with parents and siblings, in school, with friends and others, they are also gaining their first sense of who they are in the world. At the same time, at night, they are experiencing frequent nightmares. I think these two aspects of their lives are intertwined.

Here's why. As we explored in Chapter 1, we are not born able to dream; our ability to dream develops during childhood. The dreams of children and their waking imagination grow in tandem. As children develop visual-spatial skills that make them capable of imagining a three-dimensional world, dreams begin to resemble video, rather than stills. When children reach the age of five, they start appearing as figures in their dreams and characters in the dream plot. This is a normal phase of development, no different from learning how to crawl or walk, or how to ride a bike. And this is when nightmares begin.

One thing that gives nightmares an extra kick of terror for young children is that they can't distinguish the difference between dreams and reality. The phrase "it was only a dream" means nothing to a five-year-old. We know this thanks to extensive research of what age children are when they understand their dreams are private, that they are imagined events not witnessed by others. The arrival of a dream *self* and nightmares at the same time in a child's development is likely no coincidence. Nightmares may just be a universal cognitive process by which all children cultivate and forge a sense of themselves as independent minds—truly separate from others—and even help them differentiate dreaming thought and waking thought.

A sense of *self* is not something we think about much as adults. Our sense of *self* is fully formed. We know who we are. We have an understanding of our existence as individuals, of our character and physical features, of our thoughts and feelings, of who we are relative to others: parent, child, sibling, partner, friend, adversary, co-worker, and so on. To be human is above all to navigate a complicated social landscape. This internal and external sense of who we are is sometimes called the narrative *self* and the social *self*. As children, this is all new territory. Becoming an individual is a learning process. Young children are only beginning to understand that they have their own rich and unique inner life, and also a place in the real world, within families, tribes, towns and neighborhoods, schools, society, and culture. When they have that sense of who they are, children will likely show more independence and confidence, and a greater willingness to try and learn new things.

Now let's think about the typical nightmares of a five- or six-year-old, which frequently pit the dreamer versus a creature. When monsters attack children in their dreamscape, these young dreamers tell researchers the creatures are trying to invade their minds. Think about it: Children begin using their minds to create creatures that battle their own minds. The dreamer versus an evil other. Nowhere else in their young lives is their sense of *self* threatened like this.

As children age, the abundance of nightmares parallels the way in which their minds mature and take shape. For instance, the frequency of nightmares does not diminish until about age ten. Beginning at age twelve, girls are more likely than boys to have nightmares. Their nightmares are dominated by humans and small animals as aggressors, while boys more often find monsters and big animals in their nightmares. Research suggests socialization may play an important role in the differences, which start diminishing after adolescence.

As you might expect, friends and social drama play a larger role in the dreams of adolescents, and dreams start to become more sexual. With the cognitive maturation that comes with this stage, the frequency of nightmares diminishes. The common exceptions are people who have PTSD, and those suffering from mental illnesses. Even less common are people like Julia, whose frequent nightmares carry through into adulthood without any apparent cause. Like the nightmares of children, they flare up from the imagination but don't seriously disrupt sleep, cause problems with daytime functioning, or fear of going to sleep. These are symptoms of a condition called nightmare disorder that I'll cover in Chapter 5.

When we become adults, we still get nightmares, but they are typically much less frequent, maybe occurring once a month, and can be induced by life stresses. Children, too, can have nightmares born of anxiety and stress.

The themes of the nightmares also evolve as we transition to adulthood. The monsters of childhood nightmares no longer play a main role. Instead, our nightmares are more likely to include interpersonal conflicts and themes of failure and helplessness. They are also much more heavily populated with unfamiliar characters than normal dreams. As we've learned, however, there is one cardinal element of nightmares that carries through from childhood to adulthood: In a nightmare, whether it is the result of a monster or a sense of helplessness, the dream-*self* is what is typically under threat.

Imagined nightmares, like dreaming, are a cognitive achievement. If we look at the trajectory of nightmares within the broader landscape of dream life, it becomes clear that nightmares are the most remarkable species of dreams. Nightmares train the mind in ways lived experiences simply can't, helping to shape us and forge our egos. In other words, nightmares are likely necessary to our development.

The Neurobiology of Nightmares

In the 1950s, a pioneering brain surgeon named Wilder Penfield got an unexpected glimpse of the staying power of nightmares after developing awake brain surgery for epilepsy.[2] His electric probe triggered vivid and precise recollections from the past: a woman giving birth to a child, a man hearing his mother talking on the phone, the sound of a song on a record player. Patients described the experience as "more real than remembering." Penfield also repeatedly triggered a particular type of dream—the nightmare.

A fourteen-year-old girl reported a terrifying experience from childhood that had become a recurring nightmare. She was walking in a meadow. Her brothers had gone ahead of her on the path, and a man was following her, saying there were snakes in the bag he was carrying. She ran from the man, hoping to catch up with her brothers, a scene that replayed in her nightmares. Each time Penfield's probe touched this spot on the brain, he was able to trigger the scene, which preceded her seizures.

In the story of the awake brain surgery patient I shared at the top of the opening chapter, I, too, was mapping my patient's temporal lobe when I triggered a nightmare. Usually, removing the probe stops the nightmare. Sometimes, however, the "switch" remains on, and the nightmare persists. This happens because nightmares—like all cognition—are powered by the flow of electricity in the brain, neuron to neuron, millions and millions of times over. My electric probe had activated the current of electricity but the neuronal activity continued autonomously, like a runaway train, her self-sustaining loops replaying horror.

When this happened, I needed to quench the circuit of electricity in this particular region of the patient's brain and stop the nightmare. I did this in the most elemental way imaginable: by fighting fire with water. As Penfield would have done. I gently poured sterile, cold water directly onto the exposed cerebral cortex to quench the electrical activity and break the nightmare. The patient didn't feel the chill, but the cold water slowed the metabolism of the neurons and made it harder for them to spark their electrical potential. And with that, her nightmare stopped.

What strikes me about Penfield's experience with awake brain surgery—and my own—is how nightmares become part of the brain's neuronal architecture. Specific, terrifying scenes have taken up root in the cerebral cortex. They are encoded in a way that they can be recalled again and again with fidelity. Nightmares endure.

Science Supporting the Utility of Nightmares

Nightmares are psychologically taxing and physiologically expensive. They can quicken breathing, cause heart rates to spike, and trigger strong emotions. All this requires a lot of energy. As we keep seeing, if a trait or behavior is energetically costly—and a nightmare certainly is—it really has to earn its keep. In other words, we wouldn't be expending precious energy on nightmares if they weren't somehow useful. For this reason, nightmares can't be considered some cerebral relic, an evolutionary appendage like the appendix, once useful and now just along for the ride. Given how much we invest in them, nightmares have somehow earned their right to exist across generations of evolutionary pressures. I believe they have persisted because of their utility.

Before we learn how they might be useful, let's consider

something else about nightmares that sets them apart from dreams: Nightmares can actually be passed down from generation to generation. Researchers have found frequent nightmares cluster in families, and one study looking at more than 3,500 pairs of identical and fraternal Finnish twins found gene variants linked to nightmares.[3] If the likelihood of having nightmares can be passed genetically, can the nightmares themselves be as well? Could we be passing the scripts of classic nightmares from generation to generation? After all, outside of PTSD, most nightmares have nothing to do with daytime trauma but seem to follow well-worn scripts of terror and heart-pounding fear. A wild beast is chasing us. We're falling off a cliff. We're being attacked. Are those scripts woven into the double strands of our genetic code?

The idea is not so far-fetched. Passing beneficial behavioral traits from one generation to the next is the central tenet of evolutionary psychology, which argues that behavior is subject to the same natural selection as physical traits. For example, it's now widely accepted that genes influence cognitive abilities, such as attention and working memory. They are also said to play a significant role in traits such as a propensity for happiness or risk taking.

Another way behavioral traits learned in one generation can be passed down to the next is through something called epigenetics. Epigenetics does not change DNA, but alters which genes are turned on or off. Epigenetics allows for traits to be passed to the next generation without having to wait for glacial changes at the genetic level. In other words, the DNA doesn't need to mutate for the genetic code to be expressed differently.

There's evidence that behavioral traits, like physical ones, are subject to epigenetics. Looking at *C. elegans*, a type of worm that is a favorite among researchers: One research team found that when the roundworms learned to avoid dangerous bacteria in one generation, they passed this avoidance behavior onto the next.[4]

Humans, too, can pass traits learned in one generation to our offspring via epigenetics. The DNA in nearly every one of our cells carries a six-foot bundle of genetic code that is the entire blueprint of the human body. Cells differentiate into brain cells, skin cells, or other types of cells by picking and choosing which part of that code to copy, determining which proteins to make. Environmental changes, too, can cause different parts of the genetic code to be copied or skipped, resulting in different proteins being made. The body does this by producing molecular markers that either suppress copies of that part of the DNA or promote them.

If you're a smoker or exposed to environmental toxins, for instance, markers will appear that change how your DNA is expressed—at least temporarily. If our DNA is the blueprint for an entire house, gene expression determines whether to make a door or a window. How genes are expressed in one generation can be passed to the next, from parent to child. Stop smoking or avoid the environmental toxin and over time your DNA will return to normal.

Given that a predisposition for nightmares can be passed from parent to child, I can't help but wonder if our ancestors' dreams still ripple in some way through our sleeping minds via epigenetics.

Sleep Paralysis: The Original "Night-mare"

Imagine waking in the morning unable to move, overcome with a sense of dread, hyperventilating with terror, and feeling like a great weight is on your chest, suffocating you. You may hear a buzzing sound or feel electrical jolts or vibrations running through your body; have the sensation of floating or being touched; hear hallucinatory sounds, like laughing devils; or see a person, animal, or evil presence by you, on you, threatening you, touching you, suffocating you, or penetrating you. When this happens, you are experiencing something called sleep paralysis.

An estimate of up to 40 percent of the general population has experienced sleep paralysis at least once in their lifetime. Sleep paralysis is so ubiquitous that cultures around the world arrived at different, though strikingly similar explanations for this experience. In ancient Mesopotamia, they blamed an incubus, a male demon that wants to have sexual intercourse with a sleeping woman, or his female counterpart, a succubus. In the Abruzzo region of Italy, east of Rome, an evil witch called a *pandafeche* was thought to be responsible. In Egypt, it was a vicious spirit creature called a *jinn*. In China, a visitation by a ghost. Among the Inuit, a shamanistic attack on the dreamer's vulnerable soul. The eighteenth-century Swiss-born artist Johann Heinrich Füssli depicted sleep paralysis as a goblin-like demon perched on the chest of a sleeping woman. More recently, space aliens bent on abduction have been blamed. How else can you explain something as heart-pounding, psychedelic, and terrifying as sleep paralysis?

The word "nightmare" dates back to around 1300 CE and was originally two words: "night-mare." A mare was an evil spirit who tormented people in their sleep. Sleep paralysis can sometimes create the sensation of being sexually violated as one lies paralyzed, hence the belief that an incubus or succubus caused this terrifying experience.

One of the first published clinical descriptions of this

phenomenon was in 1644 by Dutch physician Isbrand van Diemerbroeck, who called the case report *Incubus, or the Night-Mare*. His description really captures the panic and terror of the experience: "In the night time, when she was composing herself to sleep, sometimes she believed the devil lay upon her and held her down, sometimes that she was choked by a great dog or thief lying upon her breast, so that she could hardly speak or breathe."

Sleep paralysis has two central physical characteristics: the body is paralyzed, and there is a suffocating feeling. More terrifying still, these physical sensations are typically accompanied by an ominous feeling an intruder is close by or hallucinations of a beast crouched on your chest. Let's consider how neuroscience can shed light on how each of these occur at the same time.

Paralysis during sleep is absolutely vital to keep us safe during our most vivid dreams. Otherwise, we would act them out, something seen in patients with dream enactment behavior, where a person's brain is asleep but their body is awake (something that will be expanded on in Chapter 5). As a result, they kick, thrash, and cry out in their sleep. Sleep paralysis is the flip side of the coin. The brain has awakened but the body is asleep and paralyzed. In other words, you are locked in your body.

Based on which muscles are and are not paralyzed in sleep, you can experience a feeling of suffocation and the sensation of a great weight on your chest during sleep paralysis. The diaphragm is the main muscle we use to pull air into the lungs. It is not affected by the muscle paralysis that occurs during sleep, allowing us to breathe while sleeping, but other respiratory muscles between our ribs and in our necks that allow for maximal expansion of our rib cages to pull in breath deeper into the corners of our lungs remain paralyzed. We use these "accessory muscles" when we're running up a hill—or when we're terrified by the idea of an evil presence lurking nearby. When these muscles are paralyzed, you panic. You gasp for air, but you aren't able to pull in as much as you want. I believe this creates the feeling of suffocation.

The most common element of sleep paralysis, reported across peoples and cultures, is the sensation of a lurking intruder. The likely origin of this bizarre and powerful phenomenon is the activation of a part of the brain called the temporoparietal junction, located above and behind the ears. It is the tuft of the brain that borders the temporal and parietal lobes and elicits a unique blend of phenomena when stimulated. Hyperactivity in this part of the brain will cause people with schizophrenia to attribute their own actions to others. But it is awake brain surgery that offers perhaps the strongest evidence implicating this part of the brain.

Electrical stimulation of the temporoparietal junction can induce the illusion of a shadowy figure nearby. In one case study of a 22-year-old woman undergoing awake brain surgery for epilepsy, electrical stimulation on the left temporoparietal junction produced the impression that someone was behind her.[5] The electrical stimulation was repeated twice more. Each time, the patient, who was lying down, had the sense a man was lurking. With the next electrical stimulation, the woman sat up and pulled her knees to her chest. She said the man was now holding her in his arms, and it was an unpleasant feeling. When the woman was asked to hold a card and perform a language-testing task, she said the man was trying to take the card from her. Not only was she perceiving another person in the room, she was now attributing hostile motives to his actions.

We know the temporoparietal junction uses touch and feedback so that your brain can figure out where your body is, where it ends, and where another's body begins. It seems likely that the shadow figure that is a central aspect of sleep paralysis is the result of some sort of electrical disturbance in this part of the brain, creating a creepy or malevolent "other" at the blurry edge of our imagined body.

The last part of sleep paralysis—and the hardest to explain—is the hallucinations: the goblins and devils, the incubus and succubus, the ghosts, and the space aliens we see in the space between sleep and waking. The scientific basis of this remains elusive and challenging to investigate. If I had to venture a hypothesis, it would include some sort of mismatch with the neurotransmitter serotonin coming online as we wake in a discoordinated fashion with other arousal neurotransmitters. Ultimately these hallucinations are closer to intense psychedelic experiences, which rely on serotonin modulation.

Serotonin is best known as the neurotransmitter boosted by anti-depressants known as SSRIs (selective serotonin reuptake inhibitors). Its primary function is not related to depression or mood, however, but to promote wakefulness and suppress REM sleep. During sleep, serotonin levels dip to zero. When we're awake, serotonin levels return.

Of course, sleep paralysis is not purely a physical phenomenon. Just as the dreaming brain stitches a story in search of cohesion and the comfort it provides, our brains seek to add meaning to the strange and terrible sensations experienced during sleep paralysis. Culture and beliefs play a role. This may seem hard to believe, but where a dreamer grows up and what the dreamer believes can profoundly alter the experience of sleep paralysis. If you grow up in Italy or Egypt or somewhere else where folklore blames evil witches, demons or other malevolent forces for sleep paralysis, your sleep paralysis will be different and perhaps worse than for someone who grows up in a country without these myths. Your mindset matters.

Think about it. If you wake up and you are paralyzed, your level of panic will be much greater if you think an evil being is lurking nearby or that your physical discomfort is the result of a malevolent force attacking you. In your terror, breathing may be more labored, the pressure in your chest greater, the experience more traumatic.

If you wake up with sleep paralysis, what can you do? How are you supposed to deal with the suffocating sensation, the squeezing on the chest, the hallucinations, and the fear? After all, few things in life are this terrifying. It comes down to using your waking mind to take on the surreal and terrifying signals you're getting. The mind has created the sense of fear and terror, so the mind can be used to deactivate it—to turn down the panic and the fight-or-flight fear response.

When you experience sleep paralysis, don't try to move, and remind yourself that what is happening is benign, temporary, and no real cause for fear. It's a good idea to keep your eyes closed and tell yourself that any presence you perceive in the room is imagined. Inward-focused meditation on something positive can also help.

How to Alleviate Nightmares

Let's go back to Julia for a moment, our case study from the beginning of this chapter. After a nightmare, Julia would sometimes wake up shaky or with her face wet with tears. The following day, she would be anxious, and the intense emotion of the nightmare would stay with her.

Nightmares are the most intense dreams, and the ones hardest to forget, and as a result, they can raise daytime anxiety, just as they raised Julia's. In the day following a nightmare, most people are typically more anxious and on a less stable mental footing than on nights when they did not have a nightmare. In one study, nurses who kept sleep diaries had more nightmares after a stressful day, and had more stressful days after nightmares. In other words, it's possible to enter into an unhealthy spiral of nightmares and stress.

One common response to frequent nightmares is avoiding sleep altogether. After all, you can't have a nightmare if you don't fall asleep. Unfortunately, this type of self-induced insomnia only further misaligns our circadian rhythms, resulting in more nightmares. How then can we tackle the type of recurring nightmares that are not the product of trauma?

First, we need to remember that dreams are an extraordinary act of imagination. As our most elaborate dreams, nightmares are the ultimate imaginative act. When we dream, our attention is turned inward and our Imagination Network is activated. But that doesn't mean it is operating autonomously. It is the product of our brain and affected by our state of mind. This means that we can play an active role in our dreams.

Researchers have shown a process called autosuggestion, or dream incubation, can work to steer dreams in a certain direction. This is how it works: As you go to sleep, simply make a verbal statement: "I want to dream about X." Better yet, create a mental picture of the person or question you want to dream about. You can also place a picture of who or what you want to dream about on your nightstand. Dreams are visual. Do either of these, and you are speaking the language of dreams.

Because anxiety and stress can provoke nightmares, therapy and other techniques to lower daytime angst can reduce the frequency of nightmares. A calming bedtime ritual such as meditation or yoga is also thought to help, given how dreams reflect our emotional state. If we change how we're feeling before we fall asleep, we are likely to prime our dreams.

After a lifetime of frequent, violent nightmares invading her otherwise peaceful life, Julia, at a friend's suggestion, went to see a therapist to learn about something called Imagery Rehearsal Therapy. The goal of Imagery Rehearsal Therapy is to defang a nightmare by rewriting it. Guided by a trained professional, Imagery Rehearsal Therapy is usually a two-step process consisting of four sessions, each lasting two hours. The first two sessions dig into the impact of nightmares on sleep and how nightmares may develop into a learned behavior. The second two sessions teach how to use daytime imagery and rehearsal to fundamentally alter the nightmare into something else. Imagery Rehearsal Therapy may sound too simple a solution to something as profound and piercing as nightmares. But the technique has been rigorously researched in well-designed studies that have shown benefits for others long after the therapy sessions ended.

For Julia, Imagery Rehearsal Therapy required thinking of a recurring nightmare she'd had, changing the plot from one of violence and terror into something more pleasant, happy even, and repeating that story to herself while she was awake. It turns out the way we break nightmares sheds light on their origin. Nightmares, like all dreams, are the product of our imaginations. The same imagination that is the source for nightmares can be used to break their terrible spell. We can treat wild and dark flights of imagination with a sunnier version of the same tale. With Imagery Rehearsal Therapy, the plots to the new, more benign dreams can be written or visualized, including as many specific details as possible.

Julia decided to try this technique on a particularly upsetting recurring nightmare. Like many nightmares, this one started as a normal dream. At first, she's walking in a picturesque town in southern Spain with her best friend. Then, everything changes for the worse. Bombs start falling, causing mayhem and bloodshed. Julia and her friend run in panic, trying to escape, but never find a way out. Using Imagery Rehearsal Therapy, Julia rewrote the script of her nightmare. On the advice of her therapist, she added sensory details like smell, touch, and taste. The reimagined dream started the same, a walk in a beautiful Spanish town, but instead of bombs dropping on the old houses, she and her friend hiked out of town and sat in a park redolent of flowers and punctuated by beautiful old trees, feeling a warm wind on their faces. Julia typed up the new, pleasurable version of the nightmare and read it to herself every afternoon for a couple of weeks. As she read what she'd written, she pictured the rewritten dream in vivid detail.

At first, Julia was skeptical. Could there really be such a simple solution to calm her nightmares? To her surprise, Imagery Rehearsal Therapy worked.

She tried using this method on other nightmares that plagued her. In the nightmare of a man following her on an empty street at night, he is no longer out to harm her but is trying to return something she'd lost. With the new plots, Julia's nightmares started the same way, but the endings were changed to ones that were benign or pleasant. In the four years since she undertook Imagery Rehearsal Therapy, Julia rarely has nightmares.

Ultimately, the source of Julia's nightmares was never clear. She appeared to live a life largely free of the anxiety and stress that can trigger nightmares. She did not report being depressed, nor did she talk of trauma in her life, which can cause its own nightmares. As we'll see in Chapter 5, frequent nightmares that begin in adulthood can at times signal more serious health concerns, but this did not appear to be the case with Julia, whose frequent nightmares began as a child. Julia may simply be one of those people for whom the nightmares of childhood simply never went away, a cognitive process that burned brightly in her youth and were never fully silenced.

Lucid dreaming can also work for chronic nightmare sufferers in much the same way Imagery Rehearsal Therapy can. Lucid dreaming occurs when you know you are dreaming while remaining asleep (more on this in Chapters 6 and 7). Instead of rewriting the script ahead of time, lucid dreamers can change nightmares on the fly—as they are happening. Studies have shown lucid dreaming can not only reduce the frequency of nightmares but make them less frightening. Researchers studying lucid dreaming have also found that although not everyone is able to dream lucidly during their nightmares, on the whole, the participants reported fewer nightmares and changes to the nightmares themselves. Perhaps merely the belief that nightmares can be overcome is enough to change them.

An important caveat: Mitigating recurring nightmares stemming from PTSD poses a different challenge, because these dreams do not arise from the dreamer's imagination, the way Julia's did, but from trauma. The nightmares of PTSD sufferers are essentially flashbacks in the sleeping brain. Because these nightmares are based in reality, trauma dreams can be more distressing than a typical nightmare and can't be written off as a disturbing product of the imagination. Derailing this pattern of trauma-triggered nightmares has been particularly challenging. A drug that blunts the fear and startle response has provided a partially successful approach, but the medication has common side effects that include dizziness, headaches, drowsiness, weakness, and nausea.

Barry Krakow at the University of New Mexico decided to see if Imagery Rehearsal Therapy could work to alleviate PTSD's recurring nightmares the way it does non-PTSD-induced nightmares.[6] He tried the technique with a group of sexual assault survivors who had moderate to severe PTSD. The study participants went through three, three-hour sessions. They first learned that their nightmares, which may have helped them emotionally process the trauma initially, were no longer serving a useful purpose. Then, they were told the nightmares could be targeted the way you might a habit or a learned behavior. Finally, they picked a single nightmare, were told to rewrite it any way they wished, and then rehearsed it for five to twenty minutes per day by imagining the revised dream. They were also told to avoid talking about the trauma or the content of their nightmares. Krakow and his team followed up with the research subjects at three months and six months. They found rescripting and then rehearsing these altered nightmares helped to reduce the number of nightmares and improve sleep quality.

At first glance, nightmares make no sense. They are unpleasant and don't appear to have any value for our waking life. But troubling as they are, nightmares are not a glitch. They arrive for us all as children and are deeply rooted in our neurophysiology. They cultivate our young minds in a way lived experience simply cannot, helping us define ourselves as individuals, separate from the people around us. In that way, nightmares help make the mind. And as we'll see in the following chapter on erotic dreams, the mind can also make the brain.

3.

Erotic Dreams: The Embodiment of Desire

Erotic dreams are part of human nature. You couldn't stop them if you wanted to. Menopause does not end erotic dreams, nor is chemical castration enough to extinguish them. It doesn't depend on whether you are sexually active, celibate, married, or single. Erotic dreams are universal.

In general population surveys conducted across the globe, sexual dreams were reported by 90 percent of Brits, 87 percent of Germans, 77 percent of Canadians, 70 percent of Chinese, 68 percent of Japanese, and 66 percent of Americans. If the question is framed more broadly to include all erotic dreams and not just sexual dreams, the response jumps to well over 90 percent. Nightmares are the only other type of dream that is essentially universal. Both nightmares and erotic dreams have an outsized impact on our waking lives, indicating that, perhaps, they are meant to.

An estimated one in twelve of all dreams contains sexual imagery. There's some disagreement among studies, but the most common imagery in erotic dreams appears to be, in order, kissing, intercourse, sensual embrace, oral sex, and masturbation. Kissing appearing at the top of the list should come as no surprise, considering that when you map the cerebral cortex for real estate assigned to different sensations, the tongue and lips have a disproportionately large space devoted to them.

Whether they are about kissing or something more, erotic dreams are hard to ignore. They can leave us flushed with pleasure or filled with jealousy. They are often unsettling. But what does it mean to have a sexual dream about an ex? What if your partner has an erotic dream about someone else? Should your dream or your partner's worry you? Do they reveal anything about our true desires?

Erotic Dreams Are Another Form of Imagination

Looking at how relationship status affects erotic dreams, we see single men have a higher frequency of erotic dreams compared to men in stable partnerships. On the other hand, women report more sexual dreams when they miss their partners or at the height of a love affair, while men report no similar surge in erotic dreams in those scenarios. But there's one way in which the dreaming life of men and women align: Almost all of us cheat in our dreams.

What should we make of this? As creators of our dreams, we select the cast of our nocturnal dramas, the stage, and the action. The dreams we conjure are our own sensual productions. Wouldn't a dream where we cheat on a partner be a sign we are looking to be unfaithful, or at least open to it? If an erotic dream is not our libido unfiltered and unleashed, then what could it be?

To consider this question, we first need to look at what erotic dreams are and how they are produced. As we've learned, all dreams are the product of our imagination and the Imagination Network, a visual and emotional story unbound by the rules and logic of our waking life. When we're dreaming, the Executive Network is idle and the Imagination Network is unfettered, free to find loose associations and connections in our memories and among the people in our lives. Looking at things in new ways helps us better understand our past experiences and may give us a clearer sense of what to expect in the future. The same freewheeling mindset in our dreams that allows us to explore scenarios that may be unimagined or unimaginable during our waking hours can also lead us to think about the people in our lives in surprising, disturbing, and even erotic ways.

Because the logical Executive Network is shut down during dreaming, we can't stop these erotic flights of fancy before they take off, and they are free from judgment—even our own. In our dreams, we can imagine sexual encounters and scenarios that would be taboo or inconceivable in our waking lives.

Dream reports compiled by researchers reveal just how liberated we are in our dreams. Erotic dreams are most often not a recreation of our waking sex life, and in fact, if we are in a relationship, most erotic dreams do not even involve our current partner. Instead, in our erotic dreams, we have much more of an inclination toward bisexuality and novel sexual interactions generally.

In dreams, we are free to be with anyone we want. Given this freedom, whom do we desire in our dreams? This may surprise you, but we don't conjure the ideal sexual mate to participate in our erotic dreams. We don't create some sort of idealized chimera, blending desirable characteristics into the ultimate dream fantasy. We typically imagine someone closer to home, often someone prosaic—possibly even repellent—from our waking life. That's why erotic dreams tend to involve someone familiar: our ex-partners, our bosses and co-workers, friends and neighbors, even family members when we're younger. Four out of five erotic dreams involve someone well-known to the dreamer, and these erotic encounters generally occur in a familiar place.

This also means there is such a thing as a sexual nightmare.

Dreaming about strange sex with weird or unpleasant people may be disconcerting, but it could be the Imagination Network exploring another type of social cognition, perhaps a power dynamic played out in an erotic setting.

Of course, erotic dreams can also involve celebrities and other public or historical figures. For that, we have something dubbed the Halle Berry neuron to thank. Several major academic collaborations between neurosurgeons and scientists discovered that we have individual neurons dedicated to people and places most familiar to us. This includes family members, our childhood home, and famous people and places. We might have a neuron that lights up for the Sydney Opera House or the Eiffel Tower, for instance. In the same way, we also have neurons that light up for certain celebrities.

Professor Rodrigo Quian Quiroga of the University of Leicester in England made this startling discovery by studying patients who had hair-thin wires called electrodes inserted into the cerebral cortex of their brains in advance of surgery for epilepsy.[1] The electrodes were there to detect electrical signals. The brain is normally alive with waves of electrical activity; epilepsy disrupts that.

Think of epilepsy as an electrical storm in the brain, rogue brainwaves that overwhelm normal brain activity. Medications usually stop epilepsy, but if they don't, patients can choose surgery. But for the surgery to be effective, we need to know exactly where the seizures originate—what we call the ictal onset zone—and how they spread. The brain mapping to figure this out is done with the patients spending days to weeks in the hospital, where we monitor their brainwaves until a seizure breaks through. Once we know where the seizures start and how they move across the brain, we can stop them surgically by dissecting brain tissue at the seizures' ground zero.

In the patients Quiroga was studying, the intracranial electrodes were inserted where the seizures were thought to originate, on the medial temporal lobes, which sit just above and forward from the top of your ears, and deep toward the middle of the brain. Two key structures related to memory, the hippocampus and the amygdala, are located in the medial temporal lobes.

Quiroga wanted to see what was happening at the level of individual neurons. Using the inserted micro-electrodes and a technique called "single-cell recording," Quiroga used the signals from the intracranial electrodes to see whether individual neurons were firing. Think of it as looking at an individual wave in the ocean, rather than the tide. Previous single-cell recordings done at the University of California, Los Angeles, had shown that individual neurons that are electrically activated in the medial temporal lobe could distinguish between faces and inanimate objects, and among specific emotional expressions, such as happiness, sadness, anger, surprise, fear, and disgust.

Using single-cell recording, Quiroga was able to show something astonishing: Individual neurons responded selectively to pictures of celebrities. In one patient, a single neuron responded to pictures of American actor Halle Berry but ignored other images of people and places. It even lit up to a picture of her in costume and again to an image of just her name. In another patient, a single neuron responded to pictures of Jennifer Aniston while ignoring other famous and non-famous people, as well as pictures of animals and buildings.

This response to images of celebrities at a cellular level in the brain reveals the oversized influence celebrities have in our lives. It's fascinating. Celebrities have literally taken root in our neural architecture. Our response to them suggests they are as familiar to us as a long-time friend or neighbor. Because they inhabit a physical place, down to the level of neurons in our brains, it's reasonable to conclude that erotic dreams with celebrities constitute dreaming of people who are very familiar.

Whether they involve a neighbor or a celebrity, do erotic dreams mean anything? If so, what are they telling us?

At the heart of these questions is both the relationship between our dream persona and our true *self*, and the dreamscape's relationship to our waking world. If dreams are an accurate reflection of our waking lives, and our dream *self* is the same as our waking *self*, then whatever we do in our dreams is something we'd do—or wish we could do—when we are awake. If that were the case, if dreams are merely a continuation of our daily lives, then our dream reports and our daily diary entries would be indistinguishable.

But as we know, that's not true. So how closely are our waking *selves* and our dreaming *selves* connected? And what is it that drives our erotic dreams?

What Drives Erotic Dreaming?

For a long time, researchers have tried to connect the dots between what we do when we're awake and what might lead to an erotic dream. They've conducted surveys asking about sexual activity, how happy people are in their romantic relationships, whether they have jealous personalities, and other waking life behaviors and personal characteristics. They have even tried to provoke erotic dreams by asking study participants to watch pornography before spending the night in a sleep lab. What they found was surprising. Erotic dreams are not tied to how much sex you are having, nor are they related to whether you masturbate. They are not even connected to how much pornography one consumes. The best predictor of erotic dreaming is how much of our waking life we spend day-dreaming about erotic fantasies.

Think for a moment how provocative this is: It is not what we do during our waking hours, but what we *think about* that feeds our erotic dreams.

But why would erotic dreams be linked to daytime fantasies? Why wouldn't they be linked to actual sexual behavior instead? In thinking about these questions, it's important to remember the creative engine behind our dream narratives: the Imagination Network. If our imagination is more active when we're awake, if we are more prone to daydreaming, we are more likely to have creative dreams. Along these same lines, if our erotic imagination is more active when we're awake, it is possible that we become more open to erotic dreams at night.

However, there's one very important difference between daytime fantasies and erotic dreams. When we're fantasizing during the day, these erotic thoughts are reined in by the Executive Network, which constrains sexual desires. This moderating influence on our erotic imagination when we're awake is gone when we dream, allowing our erotic dreams to be wildly creative and exploratory.

If our daytime fantasies are visions of some desired, if improbable, sexual outcome, erotic dreams are more like prurient thought experiments. We can switch genders or become bisexual in our dreams, even if it never crosses our minds during the day or in our most liberated fantasy. This is not necessarily a sign of the type of latent desires Sigmund Freud wrote about, but it could be a kind of cognitive platform on which sexual fluidity and ingenuity evolved for the benefit of our species. As we've seen before, these kinds of "wildcards" in our dreams could help us as a species become more adaptable when the unexpected happens, the black swan events that require creativity and resilience for survival. Our wildly creative and adventurous erotic dreams give us a plasticity of desire and prepare us to always procreate. If half the tribe is wiped out by disease or killed, erotic dreams like these could have readied our ancestors for new engagements and entanglements within our tribe. This may also help explain why erotic dreams tend not to look outside "the tribe" but stick close to home. The characters in our erotic dreams are rarely inventive but the interactions often are.

In this way, erotic dreams are more than our true desires: They are the embodiment of desire itself. Erotic dreams prime us for sexual exploration and the plasticity of attraction with a breadth of sexual impulses. This makes sense when we remember the essential biological imperative of life is to survive at least long enough to procreate.

Erotic Dreams Arrive before Erotic Behavior

As a twelve-year-old seventh grader, Izzy started keeping a dream diary when she started having sexual and romantic dreams about a celebrity. She kept the journal until she was twenty-two, more than 4,300 dreams in all, and she donated it to an online repository for dreams and dream research called the DreamBank.[2] In addition to sometimes dreaming about family members, Izzy's diary details a long succession of dreams based on crushes on classmates or actors. In one dream when she's thirteen, she's a boy having sex with one of her girlfriends. At age seventeen, she dreamed she was intimate with a man as part of a movie scene.

What's interesting is that Izzy told researchers she had not been involved in any sexual activity until she was twenty-five. Why would Izzy experience erotic dreams ahead of any first-hand experience? Could it be that her erotic dreams might have been the cognitive cue for reshaping the brain ahead of sexual activity?

To answer these questions, we should first know the difference between the brain and the mind. When I mention the brain, I mean its physical structure: the various lobes that comprise our four pounds of thinking flesh. The mind is different. The mind refers to what emerges from the brain's physical structure. This includes the connections and coordination between the lobes, how the neurons fire, and so on. The brain could be a city map showing streets and buildings, the power grid, and subway tunnels. The mind is the movement of people and vehicles engaged in different activities that constitute life. Or think of it as a computer. If the physical brain is the hardware, the mind is the software. Unlike a computer, though, where software is created externally and downloaded onto the hard drive, with the mind and brain the two are intertwined and inseparable. The brain creates the mind, but the mind can also alter the brain from which it arose. There is a reciprocal relationship. The mind both shapes the brain and emerges from the brain.

Given this relationship, it seems erotic dreams might not just reveal what we are experiencing, but drive what we *must* experience. Let's explore this more closely.

At birth, our brains are a starter kit, a work in progress that requires experience and learning to develop. We're born with more neurons than we have as adults and only keep those that are useful. Experience prunes the brain cells that sit idle while expanding the branching connections between the neurons we do use. If a child takes piano lessons, the brain will change in the areas involved in playing an instrument, mostly the cortical motor system, but also in the auditory system and the corpus callosum, which connects the two halves of the brain. In other words, the parts of the brain that are used thrive, while others wither. Use it or lose it is the general rule.

In our brain, each of our five senses lands on its corresponding region on the thin outer layer of neural tissue, called the cortex. Hearing lands on the auditory cortex, taste on the gustatory cortex, smell in the olfactory region, vision in the occipital cortex, and for touch, there is the sensory cortex. But we also develop another type of sensation from a lesser-known part of the brain that blossoms in adolescence: the genital cortex. This is an extension of our sensory cortex, which is an undulating series of ridges of the brain that goes from the top of our ears to the top of our skull.

The genital cortex is the sensory representation and map of our sexual organs on the surface of our brain. It has a specific address in each of us, is identical in men and women, and is consistently detectable on the topographical map of our brain. I believe, when it comes to how much we can get turned on, we are all created equal.

The tiniest electrical jolt to a spot on the genital cortex can trigger sexual thoughts. For example, one patient whose brain was mapped with electrical stimulation in the regions near and on the genital cortex in the presence of several researchers said, "That felt good; quite erotic. I can't explain it."

As recent studies have shown, it's not just genitals that

can send signals to the genital cortex. Many other areas are potentially erogenous: nipples, chest, parts of our back, thighs, and even toes. Therefore, a more accurate name for this cortex might be "erogenous cortex" or "erotic cortex," leaving the door open for any and all sensuality to arrive—from any touch, anywhere, based on the intention and the perception.

This unusual sequence of neurodevelopmental events—sexual dreams, followed by the brain's expansion of a region that allows touch to become erotic, all before we engage in sexuality—is highly suggestive that the mind may actually create and cultivate the brain. Increasingly, we are able to demonstrate that our thoughts and emotions can in some cases sculpt our brains through a process called activity-dependent myelination.

When we think a certain way repeatedly, or when we behave a certain way habitually, those neuronal circuits in our brain attempt to become more efficient by wrapping the neuronal extensions, called axons, with insulation, which is called myelin—think of the rubber sleeves covering wires at home. Myelin is made of a particular fat, omega-3, a so-called good fat that allows electrical currents to run faster and spread faster. Activity-dependent myelination is a fundamental process of how the mind can alter the very structure from which it arises.

Based on the sequence of events—erotic dreams followed by the arrival of erotic touch—an elegant hypothesis would be that erotic dreams shape the brain by promoting the creation of the genital cortex during prepubescence. With the genital cortex in place, erogenous touch becomes possible. And with eroticism possible, a cascade of hormones activates our bodily sexual maturation.

The Brain Is Our Most Powerful Sex Organ

Erotic dreams are undeniably deeply pleasurable. A survey of university students in China[3] received an overwhelmingly positive response to the following statements:

- I sometimes hope to immerse myself in a sexual dream and never wake up
- I feel lucky to have sexual dreams
- I feel sadness after waking up from a sexual dream because I find that it was just a dream
- After waking up from a sexual dream I try to continue it in my imagination

How can it be that imagined sex carries such emotional, libidinal weight? These are, after all, solitary, imagined events outside of our conscious control. It seems implausible they could mean so much to us, but they do.

There might only be one possible answer: Erotic dreams have this kind of power because the brain is our most powerful sex organ.

Erotic dreams do more than reflect or release our emotions, imagination, and libido. They can deliver the same intense pleasure as actual sex. I could argue that they may be in some ways better than the real thing. Let's consider the neuroanatomy of erotic dreaming.

First, though, a point of clarification: Both men and women become physically aroused when they're dreaming. Physical arousal during sleep is uncoupled from the dreams themselves. The body can be aroused even when the mind isn't. Even infants exhibit anatomical engorgement during sleep. No one is quite sure why.

In erotic dreams, the brain is not receiving any signals of touching or being touched. Erotic dreams occur in the brain alone. Even so, more than two-thirds of men and more than a third of women say they've experienced orgasms simply as the result of a dream.

What's happening in our dreaming mind that gives erotic dreams their sexual potency? To get to the answer, we need to flip the question on its head. What is happening in the brain during the physical act of sex?

Sensual and sexual activity draw upon every bit of neural fiber throughout our nervous system: the central nervous system, which is the brain and spinal cord; the peripheral nervous system, which are the nerves that leave our spinal cord and reach every millimeter of our skin; and the autonomic nervous system. The autonomic nervous system is often referred to as "automatic" because it can function outside of our conscious intentions. It covers our viscera, meaning lungs, abdomen, and pelvis, and has both the sympathetic nerves—capable of triggering the fight-or-flight response and peppering our tissues with adrenaline, boosting our heart rate, and putting the digestive system on pause—and the parasympathetic nerves, which return our heart rate and our guts to normal. This is the rest and relaxation counterbalance to fight-or-flight response. The distribution of the autonomic system is primarily in our core, stomach, chest, and pelvis. This could be why orgasms feel so visceral, expansive, and deep.

The peripheral, sympathetic, and parasympathetic nervous systems all send signals to the brain during sex. More importantly, the brain interprets them. Consider a simple touch. You can be touched in the same place, with the same pressure, in the same fashion, and your brain can neglect it as something insignificant, or interpret it as a frisson or a caress. It doesn't matter where you're touched. Touch can turn erotic anywhere on the body. The brain alone is what determines sexual salience, causes us to feel attraction (or not), our breathing to quicken (or not), our heart to race (or not), and for us to become aroused (or not).

During sex, our egg-shaped thalamus in the deep center of the brain relays sexual cues coming from peripheral nerves via the spinal cord. The medial prefrontal cortex (mPFC), the newest part of the Imagination Network involved with social cognition and stitching together stories, categorizes those erotic stimuli and liberates them with its imaginative prowess by adding fantasy to the experience. The amygdala, which is responsible for our instinctive fear response, also imprints emotional significance to any experience, including sex.

Now, let's return to the erotic dream. In erotic dreams, the body is silent. The peripheral and autonomic nervous systems are not sending signals to the brain. Remember, during our most vivid dreams, our autonomic nervous system is accessible, but our muscles for coordinated movement are essentially paralyzed below the neck. The brain is not reacting to any signals from the body but acting out its own imagination. There is nothing to interpret. We think of the body and brain to be an extension of one another, and in many ways it is. But in dreaming, the brain can and mostly does act autonomously.

As erotic dreams show us, the brain doesn't need the rest of the body at all. Even without signals from the body, the brain can create its own stage, characters, and action. The mind is its own erogenous zone, and dreams can pursue the pleasures of the flesh without any flesh other than the brain itself. This is another example of stimulus-independent cognition.

If this all sounds impossible, think about other aspects of

how we perceive and respond to the world. Consider sight, for example. When we're awake, we take in the visual world with our eyes. The lens and cornea work together to focus light on the retina at the back of the eye, and objects are mirrored at the back of the eye. Left is right and right is left. The perspective of each eye is also slightly different, which you can see when you close one eye and then close the other. These two mirrored and slightly different perspectives are processed in the brain by the visual cortex into a single, clear view of the world. Without the brain, we do not see.

Erotic dreams are much the same. With no sensory inputs at all, the brain creates and perceives full-bodied pleasure. Sex and other erotic pleasures we experience in our dreams are not felt any differently, because as far as the brain is concerned, there is no difference. The brain does not experience real orgasms or fake ones; to the brain, they are all real. And since the level of activation of the emotional, limbic system when we're dreaming can exceed levels we reach in waking life, it's reasonable to conclude a dream orgasm can take us to emotional heights waking sex cannot.

What Erotic Dreams Reveal about Our Relationships

Based on the neuroscience and dream diaries, dreams of infidelity are unlikely to be a signal we want to be unfaithful and are much more likely to be the Imagination Network in action. Cheating on a partner in a dream may simply be a sign of curiosity and normal sexual arousal, rather than a desire to stray from the relationship.

Nor are dreams where we explore a different sexual orientation a sign of a secret or repressed desire. This, too, appears to be more curiosity, libido, and imagination at play, or our brain's way of preparing the species for procreation.

Even so, erotic dreams have plenty to tell us about both the health of our current romantic relationships and how well we have gotten over former romantic partners, but perhaps not in the way we may expect. Erotic dreams can elicit strong feelings of desire, jealousy, love, sadness, or joy powerful enough to affect how we feel about our partner the next day. Just like the sensations in the dream, the brain perceives the emotions as real. Researchers have found conflict with a partner in a dream tends to result in conflict the following day.

In unhealthy relationships, infidelity dreams are associated with decreased feelings of love and intimacy in the days that follow. I need to underscore that the decreased feelings of love and intimacy occurred only in unhealthy relationships, not healthy ones. In healthy relationships, infidelity dreams don't have much of an effect at all.

How we feel about a partner during our waking hours can also affect our dreams. Feelings of jealousy during the day can produce dreams of infidelity, which in turn affect a dreamer's behavior toward their partner. In these cases, dreams and reality appear to feed on each other in a negative loop.

A questionnaire completed by undergraduate students[4] found they were more likely to be unfaithful in their dreams if they were romantically jealous of their partner, and less likely to be intimate the following day as a result. They were also more likely to have dreams of their partner cheating on them if infidelity had happened in real life. Based on the research, it's likely that negative emotions in an erotic dream about a partner could serve as an important signal of how you feel about that person. The emotions associated with erotic dreams are far more important than the dream narrative itself. Because of the hyperactivation of our emotional, limbic system structures that drive emotions, this is true of dreams in general and offers one key signpost to finding the meaning in your dreams (more on this in Chapter 9).

At the end of the day, is having an erotic dream about a current partner a good sign? The answer appears to be: It depends. If a relationship is going well, having a sex dream about a partner is likely to produce more intimacy the next day. If things are not going well, sex dreams are associated with less intimacy the next day. Why this is the case is not entirely clear, though it may be that the dissonance between the erotic dream and the troubled relationship results in a greater sense of dissatisfaction.

If you or your partner have a dream of being unfaithful, this is not a sign of anyone's true desires. Even though you may wake up unsettled or upset, remember that dreams are designed to make us think divergently, including about our sex lives. While healthy relationships do appear to buffer the negative effects of infidelity dreams, what really counts is not our erotic dream narrative or our partner's but how we react to these dreams.

Erotic dreams can give us clues not only about our current relationship, but about past relationships as well. Ex-partners can and do show up in dreams long after they have ceased being a part of our lives. A DreamBank contributor named Barb Sanders dreamed about her ex-husband about 5 percent of the time, twenty years after her divorce.[5] While dreams of current partners often involve doing something together, dreams of ex-partners are more likely to be erotic dreams. You may be tempted to conclude this means we're longing for an ex. But based on a number of studies, the opposite is usually true. These dreams appear to be helping us to get over our former partners.

When considering what it means to have erotic dreams of former partners, it's worth restating that the emotional response your dream provokes is at least as important, and maybe more, than the action in the dream. Dreams—even erotic dreams—may simply be a way of processing the emotions of a breakup. I'll go into this in more detail in Chapter 5.

When we consider erotic dreams, we can't forget that dreams overall lean heavily toward the emotional, social, visual, and irrational. They are the Imagination Network looking far beyond the ordinary or acceptable. And though the plots of erotic dreams are often unlikely or even undesirable, the emotions behind them may offer important clues to the state of a current or past relationship.

Looking past the relational to the biological, our brains have developed so they are highly tuned to erotic thinking. Fantasy, erotic dreaming—and ultimately our sexuality—arose from the essential drive to procreate. But they flourish beyond the act of sex, to plumb the depths of emotion, arousal, and desire that only our minds can conjure.

4.
Dreaming and Creativity: How Dreams Unlock the Creative Within

A patient I'll call Anna once came to see me because a doctor had told her she had "water on the brain." This was an interesting—and inaccurate—way to describe what was going on. Cerebrospinal fluid doesn't sit on the brain. It fills the brain, surrounds the brain, and comes from within our brains in large chambers like underwater caves called ventricles.

There is a common misconception that the brain is a solid mass of cerebral tissue. It isn't. We have four large cerebral ventricles deep in the brain and narrow tunnel-like structures linking them called foramina. The ventricles produce cerebrospinal fluid. This seemingly inert brain fluid is actually teeming with invisible life—ions, chemicals, proteins, and neurotransmitters that are in a way our mind's primordial soup. The fluid nourishes and cleanses and serves as a vital buffer. If the brain so much as touched the inner bony surface of the skull, it would damage the delicate electrified tissue.

The cerebrospinal fluid is supposed to be created and drained from the brain in equal measure, so the total volume remains the same. But sometimes the fluid fails to escape in the exact amount it's made, and the surplus becomes trapped within the inflexible skull. When Anna described the fluid on her brain, she was really referring to a sequestered collection of fluid, a liquid-filled bubble essentially, that slowly formed and expanded in the slender space between the inside of her skull and the surface of her brain. A few more drops accumulated every few months for years until the bubble had grown to the size of a peach. What she had was an arachnoid cyst, given that name because the bubble is held together with a translucent membrane made of wispy gossamer cells that give it a spider web-like patina. The arachnoid cyst and the brain were competing for the same space. As a result, the contents of her skull were getting crowded.

Drop by drop, the cyst filled and expanded. Since Anna's skull wasn't about to bow outward, her brain was forced to accommodate the slowly expanding cyst. The cyst resulted in progressively painful headaches as it knuckled glacially into Anna's brain, just behind the upper and outer part of her forehead, above her right eye. This is exactly where one finds a small but extraordinarily crucial part of the brain called the dorsolateral prefrontal cortex (dPFC). This is the part of the cerebral cortex that serves as the conductor for the Executive Network. The pressure on the dPFC didn't put Anna's Executive Network out of commission, but stunned it, slowed it down, resulting in a surprising change.

Anna had always wanted to be a screenwriter and storyteller, but she had never been able to create interesting characters or nuanced stories. This had been a source of both frustration and profound disappointment. But as the cyst in her brain grew, Anna experienced an almost insatiable urge to write, what must have felt like the opposite of writer's block. Before the cyst, writing felt forced. Now it felt like a compulsion, something that gave her anxiety if she didn't get the words out.

As we chatted, I understood what was happening in her brain when she said the "volume" of new characters and storylines had just exploded in her head: Anna's arachnoid cyst had unleashed her creativity.

How Dreams Make Us More Creative

What the cyst was doing to Anna's brain is a lot like what the brain does automatically when we dream. As we learned in previous chapters, the Imagination Network steers the dreaming brain toward exploring social relationships and emotion in a way that is impossible when we are oriented toward a task. This type of free thinking, focused on emotion and interpersonal drama, is also the heart of creative writing. Anna's arachnoid cyst dampened her Executive Network while she was awake, dialing down the strictures of order and reason, and giving her creative mind space to take wing. In this way, during her waking hours, she was able to think and create the way most of us do when we're dreaming.

The Imagination Network facilitates the superpower of dreaming, identifying and evaluating weaker associations in our memories, connecting dots in new, unexpected, and often illogical ways. These weaker associations are dampened during the day by design. These are the longshots, the unlikely scenarios, the implausible events that are not worth your time. If creativity and eccentricity go hand in hand, dreams deliver eccentricity, no matter how drab and humdrum we are by day. At night, when we're dreaming, the far-fetched associations played out in our dreams may unearth a nugget of gold deep in the mud. Maybe it's the unexpected answer to a problem we've been struggling with, or perhaps a new insight into a work relationship or a lover.

The creative process is a lot like dreaming. Creative thinking means approaching problems in new ways, viewing the world from new perspectives, finding connections we hadn't seen before, and coming up with solutions that have previously escaped us. Researchers call this divergent thinking, and they see this as a key to creativity. Of course, divergent thinking is not the same as creativity. Thinking differently does not necessarily lead to a creative solution or a brilliant idea. But divergent thinking is, by definition, unconventional. The alternative—convergent thinking—is focused on finding the single, correct solution to a problem. Convergent thinking might be good to fix a car, but it wouldn't be good to design one.

Now let's review how the brain approaches problems. If we're engaged in goal-directed thought, focused on a particular subject, or working on a task, the Executive Network runs the show. If we take a break, the Imagination Network turns our attention inward and lets your mind wander without purpose, just as it does in our dreams. We could be in the shower, folding laundry, walking on a familiar path, or driving on a long stretch of a boring road. When we're not actively engaged in a task, our mind is free to wander.

We do not have to think about letting our minds wander. In fact, mind wandering happens naturally when our minds are not occupied with a task, and it is thought to take up almost half of our waking life. Without any particular focus, this is often the time when creative ideas emerge. Mind wandering is conducive to those "aha" moments—unbidden insights or answers to questions we hadn't even asked. Now that we are constantly checking our phones, the times during the day when these moments are possible are becoming increasingly rare. (Take this as an open invitation to spend part of each day doing absolutely nothing.)

The insights that can arise when our Imagination Network is activated are different from logical problem-solving. Because the logical part of the brain, the Executive Network, is offline while dreaming, dreams will not directly give you the result of a math problem, nor are they likely to solve a riddle. But dreams are highly visual, so when they do deliver the answer to a problem, it's often in a visual way.

In the 1970s, William Dement, a pioneer of sleep research, gave 500 undergraduates "brain-teaser" problems to spend exactly fifteen minutes on before going to sleep.[1] He then asked them to record their dreams. Of 1,148 attempts, only ninety-four dreams addressed the problems, and only seven participants reported dreams that actually solved the problems. But when the dreams did defy the odds and solve the riddle, they were visual.

In one brain teaser, Dement told students the letters O, T, T, F, F formed the beginnings of an infinite sequence and asked them to find a simple rule for determining any or all successive letters. One of his students recounted a dream when he was walking through the gallery in a museum. He began to count the paintings. The sixth and seventh paintings had been ripped from their frames. He stared at the empty frames with the feeling he was about to solve the riddle. It was then he realized the sixth and seventh spaces were the answer. The sequence was the first letter of each number—One, Two, Three, and so on. The next numbers in the sequence were six and seven, meaning the answer for the next two letters was S. In another brain teaser, students were asked what one word the following sequence represented: HIJKLMNO. The answer is H to O (H2O), or water. One student had dreams of water but guessed the answer was "alphabet," showing that sometimes our dreaming mind is smarter than our waking mind.

Ultimately, the strength of dreams lies not in cracking logical riddles like these, but in divergent thinking, particularly when it can be represented visually. Perhaps no one has studied dreams and creativity more than Harvard psychologist Deirdre Barrett. Barrett says dreaming can free us from a preconceived idea that the solution to a problem needs to happen in one particular way. Instead, dreaming allows us to explore off-the-wall ideas we'd dismiss out of hand when we're awake, inspiration that has led to many momentous insights. Dreams have led to the discovery of the periodic table of elements, the double-helix structure of DNA, and the sewing machine, to name just a few examples.

In the early 1900s, the German pharmacologist Otto Loewi believed that the primary means of nerve cell communication was both chemical and electrical, but had not yet proven this hypothesis. Seventeen years later, he had a dream that woke him up, and he jotted a drawing on a slip of paper. In the morning, he looked at his notes but could not decipher the scrawl. The next night, the idea returned. It was the design for an experiment. "I got up, went to the laboratory immediately and performed a single experiment on a frog's heart according to the nocturnal design," Loewi recalled. In 1938, Loewi received a Nobel prize in medicine for his work on the chemical transmission of nerve impulses. This was the first evidence that nerves communicate with each other via chemicals—with what we now call neurotransmitters.

Divergent thinking can also help us look at our social interactions in new ways. Given that the heart of storytelling is the relationship between people, it should come as no surprise that researchers who have compared people with creative jobs in the film industry to the average dreamer have concluded that those with creative jobs were more likely to remember their dreams, and more likely to ascribe meaning to them. And dreams themselves have served as frequent inspiration for movie directors, who have shot scenes that first appeared to them in dreams.

More than their power to inspire, is it possible the nature of dreams themselves—with their rapid shifts in time, place and character—have also inspired the structure of stories in books and movies? Perhaps we accept the flashback, the jump from one location to another, from one character to another, because we have all experienced this narrative form in our dreams. Maybe dreams don't just lead to creativity, they provide the form for creativity itself.

Clever Ideas + Action = Creativity

In the 1800s, the structure of the chemical benzene baffled chemists. To know why, you first have to understand that carbon usually forms four bonds. For example, a lone carbon could attach to four hydrogen molecules to form methane. But benzene defied these expectations. Benzene has six carbons and only six hydrogens. If what chemists knew about benzene was correct, it should have had at least twice that many hydrogens.

The answer eluded chemists for years until German chemist August Kekulé came upon the answer—in a dream. Kekulé dreamed of a snake eating its own tail, and this led him to the solution: Benzene is a hexagonal ring. In this configuration, the carbon molecules bonded with each other and therefore needed fewer hydrogens to be complete and stable. Once the structure of benzene was known, chemists were able to use it as a building block to make everything from paints and artificial vanilla to the pain reliever ibuprofen. Kekulé's dream did not provide the solution outright, but it gave him a visual clue to follow.

As the story of Kekulé's dream illustrates, coming up with novel ideas is only half of the equation. Once Kekulé had the idea that benzene might be shaped like a ring, he had to figure out how a ring-shaped molecule would work. In the same way, the big idea is not the end but the beginning of creativity. Ideas need to be followed by action. When they're not, even the best ideas will remain unrealized. Creative juices need to be distilled, given shape, and packaged, and there's a neurotransmitter that helps us do just that.

Adrenaline is both a neurotransmitter and the hormone responsible for the fight-or-flight response. In the body, the hormone is released by the adrenal glands, which sit on the kidneys, and causes us to breathe faster and more deeply, the heart to beat faster, and blood to be diverted to the muscles. In the brain, it is a chemical messenger made from dopamine that is necessary for sorting through stimuli to find the relevant and ignore the irrelevant—distinguishing the signal from the noise, the salient fact from the chaos. Increases in adrenaline in the brain are associated with increases in cognitive performance. As levels become depleted in the brain, the opposite is true. We will have more difficulty discerning the signal from the noise and our mental sharpness will decrease. As a result, we will be more likely to select irrelevant stimuli and ignore what is relevant. In humans' distant past, when we lived a lot closer to nature and were nowhere near the top of the food chain, this type of miscalculation could be fatal.

When we're dreaming, adrenaline levels drop to zero, which allows for bizarre associations within the safety of our sleeping and physically paralyzed bodies. We do not need to pick the signal from the noise, nor can we. Because my patient Anna's Executive Network was merely dialed down and not completely shuttered by the arachnoid cyst, as it is when we're dreaming, she still had some adrenaline circulating in her brain. The level wasn't so much that it stymied the flood of characters and ideas, but it was enough that she could pick from them to shape her stories. This was the sweet spot for waking creativity.

Creativity is more than an original idea or outside-the-box thinking. It requires a base of specialized knowledge to build on and the executive decision-making to act on the idea. Anna would not have been able to act on her burst of characters and plots if she didn't know the structure a screenplay was supposed to take or remained stuck in a state of dreamy mind-wandering. Creativity is a back-and-forth process between inspiration and evaluation, ideation, and execution.

A brain imaging study of poetry composition demonstrated this beautifully.[2] The brain adeptly dialed up and dialed down the Executive Network depending on whether the poetry was being written or revised. It didn't matter if the poet was a novice or an expert. During the writing, which in poetry is highly symbolic and metaphorical, the Executive Network was dialed down. During the revision process, the network was reactivated.

The Power of Napping for Generating Ideas

Beyond the transition from wakefulness to sleep, naps between thirty and sixty minutes can restore minds worn down by a repetitive task. Longer naps, lasting sixty to ninety minutes and incorporating REM sleep, can not only significantly improve performance on a task but actually boost learning. Researchers have found that naps can also be used for creative problem solving, specifically the kind of problems that require a moment of creative insight when the answer becomes clear.

When we solve a problem with a flash of creative thinking, typically there's a gap between when we are first presented with the problem and when we figure out a solution, in which we try unsuccessfully to solve the problem and set it aside. This period when we've seen the problem but are not actively working on it is called the incubation period. We haven't forgotten the problem, but we're not actively trying to solve it, either.

Denise Cai and a research team at the University of California San Diego decided to test whether napping during this incubation period would result in better creative problem solving.[3] She divided test subjects into those who rested quietly, those who napped, and those who napped long enough to have a period of REM sleep, when we experience our most vivid dreams. Cai found an incubation period helped all three groups equally.

Cai tried the test again after the subjects were primed with clues they could use later and found something interesting. In the morning, the test subjects completed a set of analogies. For example, CHIPS: SALTY; CANDY: S___. Half the answers, SWEET, were also the answers in the afternoon word test, which was a little different. In the afternoon, they were given three seemingly unrelated words and had to find a fourth word that connected them. For example: HEART, SIXTEEN, COOKIES. Answer: SWEET.

After the associative networks in the brain were primed in this way, the group that rested quietly and the basic napping group did about the same at solving the word puzzles, but the group that napped and experienced REM sleep did 40 percent better than the other two groups. It didn't matter if this group remembered their dreams or not. They still received the creative benefit of rich, dreaming sleep.

Cai concluded that the neurotransmitters activated when the Executive Network was functioning inhibited the kind of mental associations needed to solve the puzzle. During REM sleep, however, the Imagination Network wove the new information together with past experience to create a richer web of associations. As Cai concluded, "Fluid interpretation is the hallmark of the creative mind, from idle word play to the abstraction of shapes that led to the solving of neurochemical transmission or the structure of the benzene ring."

Dreams Influence Culture

I believe no creative aspect of dreaming is more important than its power to evaluate our social relationships. Dreaming allows us to travel back in time or into the future, picturing ourselves as children once again in the company of long-dead relatives, or imagining what shape our lives could take in ten years' time or more. This comes so easily to us that we can be forgiven for failing to acknowledge what an incredibly creative cognitive feat this is. Dreams' power to take us back to a fully realized past or imagine a future embodies three remarkable human capacities: visual imagination; "episodic" memory that directly re-experiences one's own past sights, sensations, and emotions; and its temporal opposite, mental "time travel" into an anticipated future.

Each night when we dream, we are creating emotional, character-driven dramas, social scenarios that explore an enormous range of social strategies and contingencies. If dreams among early humans offered a way to help them with contingency planning on the dangers they might face, dreams today offer the same sort of virtual role playing for finding a lover or interacting with others. Behaviors can be tried out in our dreams without taking any risks to our social capital. They also give us the ability to imagine how others see us under different circumstances.

Beyond the individual dreamer, dreams have influenced writers, artists, musicians, fashion designers, architects, athletes, dancers, inventors, and others who shape the world we live in. The British novelist Graham Greene, for instance, whose novels include *The End of the Affair* and *The Quiet American*, was said to write 500 words a day, no more, which he would read over just before bed, relying on his dreams and sleeping mind to continue the work. Greene found dreams so fascinating he even published his dream diary: *A World of My Own*. The American writer John Steinbeck, author of *The Grapes of Wrath*, had a name for this overnight problem solving: "the committee of sleep."

At only eighteen, Edward Enninful was hired as art director for *i-D*, a British magazine focused on the street style of young people. He worked there for two decades before moving to Italian *Vogue*, American *Vogue*, and *W* magazines. In 2017, at age forty-five, the Ghanaian-born Briton became the first male, first gay, first working-class, and first Black editor-in-chief of British *Vogue* in the magazine's 106-year history. Enninful credits dreams with driving his creative vision.

"Sometimes I'll be really fighting with myself and not coming up with an idea and I'll go to sleep. And then I'll wake up and I'll see all the images. I'll see the model, I'll see the location, I'll see the hair, I'll see the makeup. And for years, I thought that was cheating. [It was] my mom who said, 'That's actually a gift,'" Enninful said in a radio interview.

Recovering from eye surgery and unable to see for three weeks, Enninful said he dreamed bigger still, "in Technicolor." It was during this time recuperating he envisioned what may be his most memorable cover: Rihanna as a futuristic queen, for *W* magazine.

Because dreams are so visual, they can also promote figurative thinking—when we picture something that symbolizes something else. Just as Kekulé saw the snake eating its tail as the answer to his question about benzene, dreams can be thought of more as poetry than prose, rich in metaphor.

Maya Angelou, the American writer and civil rights activist, is said to have used her dreams, but not for creative inspiration. Instead, Angelou looked to her dreams for guidance. When she dreamed of finding a skyscraper under construction and scaling the scaffolding, she saw that as a sign her writing was going in the right direction.

Do creative people dream more or dream differently? Researchers have found that creative and imaginative people are more likely to have vivid dreams, probably because there is a unique continuity in how they experience the world. If you are one of those people prone to mind wandering, there is less of a barrier between wakefulness and dreaming, so information and ideas may pass more easily from one state to another.

Kinesthetic Creativity: Dreams and Movement

Dance and other forms of movement are a fundamental type of ingenuity that often goes underappreciated. The use of tools, needles and thread, bows and arrows, and knots also require a kinesthetic creativity. Many of humankind's foundational innovations and inventions have sprung from these kinds of creativity, which require planning, motor skills, and spatial processing, and, as a result, require the use of multiple regions in the brain.

Kinesthetic creativity begins with visualizing the movement, which comes naturally in dreaming. After all, dreams are also a visual-spatial playground.

Thinking about the prowess of early humans to thrive among creatures that were stronger and faster, it's probable that their dreams gave them ideas crucial for survival. It seems reasonable that dreams fostered creativity in movement, the procedural knowledge we accumulate in our lives—and ultimately a creative well we as a species have relied upon.

Robert A. Mason and Marcel Adam Just at Carnegie Mellon University's Center for Cognitive Imaging decided to study what was happening in the brain when people were tying knots.[4] Procedural knowledge such as tying a knot is different from knowing about a thing because it unfolds over time: Tying a knot is a series of movements in sequence. Intriguingly, such procedural memory, like tying laces, tends not to fade even in dementia.

As a surgeon in training, one of the first things you learn when you get into the operating theater is the surgeon's knot, a variation of a square knot used to firmly stitch wounds together. Before the use of electricity to singe blood vessels, for example, we used knots to tie them off so we could safely cut them. Sometimes hundreds of knots were needed. If even a single knot came unraveled, it could be disastrous. Tying knots, the movement of fingers and hands, becomes balletic when done right, as if hands have a mind of their own.

In their study of knot tying, Mason and Just used a functional

MRI (fMRI) to show brain activity of test subjects in real time. The researchers found the first step of knot tying was thinking about the process before manipulating the rope. When they asked test subjects to simply imagine tying the knot, the researchers discovered something fascinating: The neural signature was exactly the same as planning to actually tie a knot. In other words, when we dream, our neurons fire as though we are performing the act we're dreaming about. This allows for dreams to enhance our procedural knowledge, which can be useful in many areas of our lives, including dance, art, and sport. The golfer Jack Nicklaus, for instance, once credited a dream with improving his golf game by giving him a new way to grip his club.

As a brain surgeon, I try to take advantage of the creative power of my dreams. The night before a particularly challenging operation, I review images of the patient's brain and brain tumor. While falling asleep, I imagine rotating the tumor, paying particular attention to the surrounding brain tissue I must either avoid or traverse. As I'm waking up, I take a few minutes to revisit the shapes and contours of the planned surgery. This practice has served well to give me the spatial awareness of the anatomy I need to either dissect through or around. Since dreams are visual-spatial experiences, I have no doubt that this mental exercise has been replayed in some way in my dreams, further strengthening my understanding of the operation ahead, even if I don't always remember my dream in the morning.

Many experiments have shown sleep and dreams help us learn. In one experiment, participants ran around a virtual reality maze. Afterward, half of the subjects napped while the other half stayed awake. When they went back to the VR maze later, those who slept performed better than those who stayed awake. Those who not only slept but dreamed did best of all. For those who didn't sleep, daydreams about the maze didn't help.

Did those who slept and dreamed do better because they were dreaming about how to get through the maze? This would be a natural assumption, but it wasn't the case. Two of the participants dreamed about music. The other had a dream of a bat cave that was like a maze but not the maze itself. Even though the participants were not dreaming of the maze, just the act of dreaming somehow helped them consolidate memories of it. They knew the maze better because they dreamed. The correlation is clear, though how this works is not yet fully understood.

Nightmares and Creativity

In 1987, Ernest Hartmann at Tufts University School of Medicine headed an in-depth study comparing twelve lifetime nightmare sufferers with twelve vivid dreamers, and twelve people who were neither nightmare sufferers nor vivid dreamers.[5] Each of the participants underwent structured interviews, psychological tests, and other measures to gauge their personalities. The researchers found that the nightmare sufferers showed greater artistic and creative tendencies than the other groups. In other words, the same minds that can imagine evil or threatening forces in their dreams can use their fertile imaginations for creative purposes in their waking lives.

Nightmares have served as inspiration for the works of many famous writers. The world's best-known horror writer, Stephen King, fell asleep on a plane and dreamed of an insane woman who held captive and mutilated her favorite writer. The result was the book, *Misery*.

The Shining was also inspired by a dream. King and his wife were the only two guests in a mountain resort hotel as it was closing for the season. There, he had a dream that his three-year-old son was running through the halls screaming as he was being chased by a firehose. The nightmare woke him up, sweating. King recalled lighting a cigarette and looking out the window: "By the time the cigarette was done, I had the bones of the book firmly set in my mind."

And what do we make of the prehistoric cave paintings and other ancient artifacts in France and elsewhere? Many of the creatures depicted across the globe are zoomorphic, a mix of human and animal. Archaeologists have wondered if these fantastic images could have been inspired by dreams. Given that nightmares are the most remembered dreams, could these be the first depictions of nightmares in art? I'd like to think so. We could make the case that storytelling itself grew from the desire to share dreams and nightmares.

How to Prime Your Dreams for Creativity

Ancient Egyptians built sleep temples where people would sleep in the hope of inducing dreams that might cure an illness or help them with important decisions. In ancient Greece, too, people would go to special temples where they would pray for a dream that would solve a problem. The Greeks called this dream incubation. Today, research tells us dream incubation is more than an antiquated practice built on faith. There's real science behind it.

Researchers have found dreamers can influence their dreams

through nothing more than the power of suggestion. Although by no means a foolproof process, they have found that simply by stating your intention to dream about a certain person or a particular subject, you can often nudge your dreams in that direction. In this way, you may be able to prime your dreams to help spark creativity, contemplate a social dilemma, and consider a major decision. Harvard dream psychologist Deirdre Barrett asked her students to think of an emotionally relevant problem fifteen minutes before they went to sleep.[6] Fully half of them reported having dreams related to the problem.

Because dreams are so visual, visualizing the person, idea, place, or problem as you drift off increases the odds your dream incubation will be successful. As we learned in the chapter on nightmares, you can recast a recurring nightmare with Imagery Rehearsal Therapy, rewriting the dream plot to make it benign or even give it a better ending. Though that approach may sound overly simplistic, you'll recall the research shows it often works to rid people of their nightmares. Incubating your dreams, too, sounds like wishful thinking, but serious studies have backed up this approach as a way to nudge your dreams in a certain direction.

Researchers at MIT's Media Lab have been working on technology to engineer sleep and dreams in a way that maximizes your creativity (see page 97). Devices sense the sleep-entry and give you a verbal cue asking what you were thinking and record your response. As we'll see in Chapter 8, there are other ways to engineer the content of dreams, using the senses.

As we saw when we were looking at how to alleviate nightmares (see page 47), you can also write down your intention on a piece of paper and put it by your bedside, or put a picture or object related to what you're hoping to dream about next to your bed. This is more than some totemic ritual. These are all real ways people report they are able to prime their dreams. It's as though we're putting the raw materials into a pot and waiting for our dreams to mix them in new and unexpected ways.

Dream incubation is most successful when the solution can be thought of visually. That's because the visual cortex is highly active during REM dreams. Before you go to sleep, review the problem or topic you'd like to dream about. Imagine yourself dreaming about the problem, waking up, and writing down your dream on the paper you have by your bed.

Barrett's students chose issues that were academic, medical, and personal, and noted which dreams offered potential solutions to their problems. In one, a student who had moved to a smaller apartment and couldn't find a way to arrange the furniture in a way that didn't seem cluttered dreamed of moving the chest of drawers to the living room. The student tried it for real, and it worked. In another, a student trying to decide between an academic program in Massachusetts or elsewhere is in a plane that needs to make an emergency landing. The pilot says it's too dangerous to land in Massachusetts. Thinking about the dream, the dreamer sees the wisdom in enrolling in a program elsewhere.

Even when you don't remember your dreams, they can influence your waking thoughts. You may have a flash of insight, an idea that pops into your head, a solution to a problem that seems to come from nowhere. The source of this spark may well be a dream, even if you don't remember it. We dream every night, and every night our dreams are putting in creative work for us.

Tapping into Our Dreams' Creative Potential

Many people do not think they are creative by nature. You might be one of them. But let's not forget dreaming itself is a creative act we all participate in. Even blind people dream, making up for their lack of visual content by experiencing more sound, touch, taste and smell than sighted people. Thankfully, the power to dream creatively is something we can all cultivate.

In our dreams, we produce compelling narratives from distant memories, recent and planned events, emotions, snippets of things we've seen online or read in a book, and other bits and pieces from our lives that we stitch together into a story. Little is off limits. The characters in our dramas can be family members, dead relatives, historical figures, friends, colleagues from work, strangers, or people we've encountered only briefly. Charlie Kaufman, the American screenwriter, said, "Your brain is wired to turn emotional states into movies. Your dreams are very well written . . . People turn anxieties, crises, and longing, love, regret, and guilt into full, rich stories in their dreams."[7]

But how can we access our creative potential? What can we do to cultivate the creativity of our dreams and direct them in a way that is most productive? The first step is to remember our dreams.

Most of us have the experience of trying to recover a dream, only to have it slip away, at first indistinct and just out of reach and then fading to nothing as it sinks back into the ocean of sleep, leaving only the faintest residue on the surface. There's a reason for this. We need to keep the boundaries between our dream *self* and our waking *self* separate. The story of our lifetime—our narrative *self*—is constructed from our autobiographical memories. These, of course, form during our waking hours, and we use these autobiographical memories to stitch together the past and project into the future. If memories of dreams were intermingled with our autobiographical memories, it would be incredibly confusing. And so the wilderness of our dream lives, where we fully inhabit and embody the experience of our dreams, is forced underground when autobiographical memory returns to us every morning.

There is one simple thing we can do to remember our dreams: State your intention. "I will dream. I will remember my dream, and I will write it down." You don't have to use those exact words, but something along those lines can work. Study after study has shown this type of autosuggestion improves our chances of remembering our dreams. There is no biological mechanism we can point to why, but it is highly probable that since some of our waking life feeds our dreaming life, this autosuggestion is retained by the dreaming mind.

When you wake up, remain still for a moment, then write down everything you can remember about your dream on a piece of paper you keep at your bedside, or on the Notes app on your phone. Don't turn on the lights. Don't check your notifications. You've got a minute or two. The goal is to delay the Executive Network from abruptly coming back online. Practice the habit of rising slowly, trying to remember your dreams before doing anything else. This type of recall can be cultivated and enhanced with effort and practice. Your ability to remember dreams will improve quickly, from a few fragments on the first morning to rich narratives in only a week or so. Whatever you do, try to write down your dreams before thinking of the day ahead.

The reality is, we forget dreams by design. When we wake up, the hegemony of the Executive Network returns and our autobiographical memory kicks in. This links our experience from day to day: who we are, where we are, what we need to do in the coming day. It's important that our autobiographical memory is not muddied by dreams. We need other types of memory: procedural memory for skills like riding a bike, episodic memory for specific events, faces, and names. Autobiographical memory ties it all together, our complete experience and not the disparate elements of our lives.

Underpinning this transition when we waken are the neurotransmitters serotonin, which is associated with wakefulness, and adrenaline, which is released the moment your attention turns outward and becomes goal-directed. This erases any chance of retrieving your dreams.

Focusing outward when we awaken is a powerful survival mechanism. After all, we are more or less helpless when we sleep. Becoming alert and quickly oriented would have allowed early humans to assess whether they were in danger as soon as they woke up. And, as I mentioned in Chapter 1, adrenaline is a powerful force, giving our waking selves the ability to search for the signal in the noise of daily life. When we dream, we turn this on its head. We are ignoring any signals and searching the wilds of the dreaming brain for patterns and meaning in the noise.

Even if we are often unable to recall our dreams when we wake up, there's some evidence the content of our dreams is still remembered. As we explored in Chapter 1, we appear to have a separate memory system for dreams. In this way, even the forgotten dreams appear to live on.

So if our goal is to recall our dreams, we need to circumvent our neurobiology, at least momentarily, to keep one foot in the dream world. Retreating into our dreams can expand our minds in ways impossible in lived experience. Thinking about dreaming and trying to remember your dreams can also expand your dream life, not unlike practicing a new language or other cognitive or physical skill.

Sleep-entry: Your Portal for Creativity

What if you could surf a liminal brain space that allowed you to toggle between divergent thinking and executive function? As a matter of fact, you can. When you are drifting off to sleep, you are in a sleep-entry state. We are both conscious and have dream-like, free-floating thoughts. In this state, we are ripe for creative thinking. In this way, the brain behaves in sleep-entry much like Anna's did with her cyst.

The surrealist artist Salvador Dalí recognized the fuzzy intersection of the dream world and the waking one as a rich source of creativity, and he developed a technique to take advantage of it. Dalí would sit in a chair holding a large key between his thumb and forefinger above a plate on the floor. When he nodded off, he'd drop the key on the plate, which would wake him up. He'd immediately sketch the hallucinatory vision that came to him as he was entering sleep. Dalí called this the "secret of sleeping while awake," and he used this as inspiration for art.

This blending of sleep and wakefulness can be detected on an electroencephalograph (EEG), which records brainwaves. As we enter sleep, the device shows both the brainwaves of wakefulness called alpha or "fast" waves, and the brainwaves of sleep, called theta or "slow" waves. It's that rare window where these two overlap. It's like an estuary where ocean and river meet, the salt and fresh waters mixing into something unique—a place where we can access the wild creativity of dreaming and be aware of it at the same time. Like dreaming, in this moment we are not guiding these often bizarre thoughts and images during sleep onset, but observing them. Like waking, we have access to these thoughts in real time. No wonder Dalí described sleep-entry as walking "in equilibrium on the taut and invisible wire which separates sleeping from waking."[8]

Researchers at the Paris Brain Institute decided to put Dalí's sleep-entry technique to the test.[9] Participants were given a sequence of eight numbers and told to find the ninth digit as quickly as possible. The puzzle could be solved slowly, step-by-step, or quickly, if they discovered the hidden rule to the pattern of numbers. The participants who didn't solve the puzzle were given a twenty-minute break and told to recline in a chair holding an object, just as Dalí did. When they dropped the object as they were drifting off to sleep, they were told to say what was on their mind just before it fell. Their brainwaves, eye movements, and muscles were monitored to verify whether they were awake or entering sleep or sleeping more deeply. After the twenty-minute break, the test subjects were again asked to solve the number sequence.

What the researchers found was astonishing: A single minute of sleep-onset inspired insight. The group that entered this twilight zone between wakefulness and sleep were almost three times as likely to solve the number sequence as those who remained awake. When the researchers looked more closely at what was going on, they discovered the sweet spot for creativity. Solving the puzzle was associated with intermediate levels of alpha waves, the brainwaves of wakefulness and the Executive Network. The subjects who did best were neither too alert, with high levels of alpha waves, nor did they feel pressure to sleep more deeply, which was associated with lower levels of alpha waves. This was the invisible wire Dalí sought. Interestingly, those participants who fell into a deeper sleep did worse after the incubation period than both those who remained awake and the sleep-entry group.

Researchers at the Paris Brain Institute had proven what had long been believed. Sleep-entry was indeed "a cocktail for creativity." The formula: a problem, followed by a brief incubation period, plus sleep-entry. The final step is returning to the problem.

As mentioned earlier, researchers at MIT's Media Lab are trying to use technology to take advantage of this window of creativity.[10] They have developed a high-tech "targeted dream incubation device" that mimics Dalí's technique by trying to gauge the onset of sleep. It uses a flex sensor on the middle finger to look for a drop in heart rate and shifts in electrodermal activity. Just as Dalí's hand opening and the key dropping into the metal bowl was his way of knowing he had slipped into a sleep-entry state, this device looks for a slow opening of the hand with an accompanying loss of muscle tone. Once sleep-entry is detected, the device gives sound cues designed to steer dreaming and spur creativity. Products like these are new, however, and whether or not they indeed work has yet to be proven as of this writing.

Sleep-entry may have powers beyond unfettered imagination and may also be valuable for learning. One study looked at subjects who were both novices and experts at the video game Tetris, where the goal is to quickly reorient falling colorful geometric shapes so they stack neatly. For the study, subjects played the game for seven hours over three days. As they were drifting off to sleep, the subjects were repeatedly roused and asked what they were thinking. Three-quarters of the novices reported seeing falling Tetris pieces during the sleep-entry state. Only half of the experts did. Among the experts, some saw the geometric images but from a version of the game they had played before the study. This suggests the novices were engaged in learning, while at least some of the experts appeared to be integrating their recent Tetris experience with previous times they had played the game.

Exactly how all this works remains a subject of intense research, especially since the novices had most of their Tetris visions on the second night of the study, a significant lag. Whether they were novices or experts, the similarity of their sleep-entry cognition was striking: They all reported seeing Tetris pieces falling in front of them, sometimes rotating and sometimes fitting neatly into open spaces at the bottom of the screen.

All this exciting research is a reminder to embrace sleep-entry as a unique state of mind, where we have released the strictures of the day but have not yet left them behind.

So what of my patient Anna? Anna may have experienced a burst of creativity when the cyst on her brain mimicked what happens in the dreaming brain, but as the inexorable drip of fluid continued, her headaches became more and more frequent and less and less bearable. With each drip, her brain was squeezed further, and she began to experience splintering headaches. People who have these headaches say it feels like the skull itself is cracking. There was a simple solution: Make a tiny incision behind her hairline and drill a hole the size of a coin to let out the fluid. The procedure wouldn't even leave a visible scar.

But Anna was reluctant. She didn't want to lose her creativity. She loved inventing worlds and didn't want to go back to simply inhabiting one. She turned down my offer, and that was the last time I saw her. I know there must have come a point where she was no longer able to endure the headaches, or her brain was no longer able to function with the growing presence, but—at that moment anyway—the risk of giving up the explosion of creativity was too much. And I understood that.

5.

Dreaming and Health: What Dreams Reveal About Our Wellbeing

It was the late 1990s and early in my training. I was at the end of the most legendary stretch of freeway in Los Angeles, U.S. 101, or the one-oh-one as we Angelenos call it, when I met a patient who completely changed the way I think about dreams and dreaming. Before seeing him, I hadn't given much thought to dreams, and I certainly hadn't thought much about how dreams are connected to our physical health, how there is a dream–body connection. But he made me look at all this in a completely new light.

To see this patient, I had to drive past the Hollywood sign, the entertainment industry studio lots, the massive LA General Hospital and the LA County Jail, all the way to the Veterans Affairs Medical Center. Military veterans in the United States have their own hospital system, and it was there that I met him: a 55-year-old man with new-onset nightmares. Like most of us, he had experienced the occasional nightmare as an adult, but now they were frequent. This was something new and concerning. Back then, I assumed nightmares in a combat veteran meant post-traumatic stress disorder (PTSD), but the patient insisted that wasn't the case. It had been decades since the war, and he'd never had any other symptoms.

What stood out to him, however, were the characters in his dreams. They were animals. I immediately thought that maybe this was undiagnosed schizophrenia. The cause of this profound and often disabling mental illness remains mysterious, but the symptoms have a striking consistency with dreams blending into waking hallucinations and delusions. Not only do patients with schizophrenia often see animals, but sometimes the animals talk to each other, and sometimes they talk to each other about the patient. But the animals in my patient's dreams were passive, just features of his dreamscape without directly engaging him. He was also able to carry on a conversation easily. Little of it added up to mental illness.

"Are you frightened by your dreams?" I asked him. The patient simply shook his head no. His medical exam and blood work were normal, but his friend said that more and more he yelled in his sleep and seemed to be acting out his dreams. The man had even struck his bed partner in the face while dreaming. This added up to something else: REM behavior disorder, where the body is no longer paralyzed during sleep, though dream enactment behavior (DEB) might be a better name.

Each night, our brains and bodies follow a sharply delineated repeating ninety-minute cycle of sleep. With each cycle, light sleep is followed by deep sleep, where the brainwaves are slow and rhythmic. After slow-wave sleep, the pattern changes again. The eyes start rolling under their lids and most of the muscles in the body become paralyzed. When the eyes are fluttering under the eyelids, this is known as rapid eye movement or REM sleep.

REM sleep and dreaming have often been described as synonymous, but this is inaccurate. We can dream in all stages of sleep. Dreaming is even possible without REM, but REM sleep is when the most intense and bizarre dreams occur. Unable to move during our dreams, we are a captive audience, safely locked into the theater of our dreams for a show we've created for an audience of one.

Based on research in sleep labs, where dreamers are awoken at different times, we know dreams change as the night progresses. Early-night dreams tend to include more elements from our waking life. Dreams at the end of the night are more likely to be emotional and incorporate older autobiographical memories, and it's these dreams just before we wake up that we're most likely to remember. The tenor of our dreams shifts, too: Dreams are more negative at the beginning of the night and become more positive as the night goes on.

Working with this patient, I came away with the profound realization that our dreams aren't separate from our body, and the link between the dreaming mind and health is far more intertwined than we possibly could have imagined.

Dreams Can Warn Us of Future Illness

We didn't know it then but this unique combination of symptoms—men in their fifties acting out their dreams—years later develops a type of brain disease called synucleinopathies. Not just sometimes, but almost always. For people whose dream enactment behavior has no known cause, within fourteen years of diagnosis a startling 97 percent will have Parkinson's or Lewy body dementia.

Synucleinopathies is a technical name for this family of neurodegenerative diseases characterized by the abnormal build-up of a protein called alpha synuclein. The small proteins exist naturally inside neurons and perform important regulatory roles, such as maintenance of the synapses between neurons. With synucleinopathies, the alpha synuclein protein is misfolded, and these misfolded proteins tend to aggregate, creating a sort of molecular sludge with dire consequences. They also appear to spread from cell to cell, causing more and more damage. Exactly how these malformed proteins result in DEB is not known, yet what a striking correlation.

Most clinical symptoms arrive with an underlying disease, but some can arrive before the disease even takes hold. In medicine, this kind of warning flare is called a prodrome, something that precedes the illness itself. A fever and lack of appetite can be a prodrome of an infection. But most prodromes occur hours or days before the onset of the ailment, not a decade or more as is the case with DEB and synucleinopathies.

In my patient's one seemingly unrelated condition, dreams predicted the pending decay of the brain and nerves in his body years before any symptoms or any diagnostic test. This man's dreams and their evolution were tied to his physical health in ways that remain elusive to our understanding. But we do know the power of DEB to predict a synucleinopathy is uncanny, about as good as any imaging or blood work to diagnose any condition, and few can do it with this type of certainty years in advance. With dream enactment behavior, patients report dreams that are vivid, violent, and action-packed. The plot of their dreams commonly involves an imminent physical threat to themselves or someone close to them. Published accounts of dream enactment among men in their fifties, sixties, and seventies document the mayhem: punching, kicking, wrestling, running from assailants or wild beasts. One man used a pillow to ward off an imagined pterodactyl. The prevalence of beasts has some similarities with schizophrenia, but again, these dream narratives do not bleed into the day as is the case with schizophrenia.

Dream enactment can be violent. Most of the time, people with DEB act out their dreams without getting up, but they do sometimes leap out of bed and fall or run into walls while fleeing imagined pursuers. A man who once dreamed he was wrestling an assailant wound up putting his wife in a headlock. In fact, men often dream they are protecting their wives from attackers only to awaken to find that they themselves are attacking their spouses. On the other hand, women with DEB are rare, and their dreams are less aggressive. And unlike men, when they act out their dreams, these dreams do not typically feature a confrontation with an assailant.

Since aggression is a thread that runs through dream enactment with Parkinson's and other synucleinopathies, it would be tempting to assume that these people are simply more aggressive by nature. Maybe their dreams are merely a reflection of who they really are. It turns out, the opposite is true. Researchers have found these combative dreamers actually score lower than average on a daytime aggression questionnaire. In other words, they are placid while awake and fierce when asleep. Why there is this strange disconnect between daytime personality and dreaming behavior is a mystery.

Not all enacted dreams are combative. Non-violent behaviors such as eating and drinking have been reported in the scientific literature, and so have DEB dreams that included laughing, singing, clapping, dancing, kissing, smoking, picking apples, and swimming. One man dreamed he was fishing as he sat on the edge of his bed holding an imaginary fishing rod.

Because acting out dreams and the new onset of nightmares are clinical harbingers of Parkinson's, arriving years, even decades, before the first movement symptoms of the neurodegenerative condition, paying attention to dreams and dreaming could offer physicians a rare window for truly early intervention. Since I met my patient over two decades ago, I've learned of other cases where dreams are said to have the potential to either warn us or inform us about our health.

In the mid-twentieth century, for example, Vasily Kasatkin, working at the Leningrad Neurological Institute, gathered dream reports from more than 350 patients and concluded that physical ailments affected dreaming.[1] Of the more than 1,600 dream reports gathered, 90 percent were negative, about such topics as war, fire, and injury. Interestingly, dreams of actual pain were very rare, only occurring in 3 percent of the recorded dreams. This finding, that physical pain is rare in dreams, has been backed up by other researchers since.

Kasatkin also found that dreams often appeared before clinical symptoms of the disease, though he did not give a precise percentage. Just as dream enactment behavior presages Parkinson's and other Lewy body diseases, Kasatkin's research convinced him physical ailments were foretold in unpleasant or nightmarish dreams. He cited a patient who dreamed of nausea, spoiled food, and vomiting who had gastritis, and another who dreamed rats were chewing through his abdomen who was later diagnosed with an ulcer. He believed dreams of the sick were different from other bad dreams because they continued throughout the night and seemed to have some link to the part of the body that was ailing. For example, a person with a lung ailment would have bad dreams that involved breathing. Kasatkin also documented dreams changing over the course of illness and recovery.

This type of dream–body connection is fascinating but also hard to prove since most of the patients were remembering their dreams after they were already sick. Maybe this was simply a case of confirmation bias, where patients became sick and then recalled a dream that seemed to warn them in some way. In an effort to come up with stronger scientific evidence, researchers have tried to capture dreams and then see how they are related to future health.

In one study, a group of heart patients were asked about dreams before they underwent a common heart procedure called cardiac catheterization, which uses wires to open narrowed segments in coronary arteries. Researchers followed these patients for six months after discharge from the hospital, and their health was scored on a six-point scale: cured, improved, unchanged, worse without rehospitalization, worse with rehospitalization, and death.

Remarkably, the researchers found the patients' dream narratives were linked to how they fared. Men dreaming of death and women dreaming of separation were significantly more likely to have experienced worse clinical outcomes, independent of the severity of their heart disease initially. This suggests that dreams somehow offered a clue to their prognosis. Were the dreams some signal of physical health? Did they convey the attitude of the dreamer toward sickness and healing? We don't know for sure, but the findings are intriguing and suggest some connection between our dreaming mind and our health.

Listening to dreams has even led to cancer diagnoses. One study described women who credited warning dreams with their decision to get checked for breast cancer—and were ultimately diagnosed with the disease. They described the warning dreams as more vivid, more intense, and having a sense of threat, menace, or dread. Some of the dreams even contained the words "breast cancer" or "tumor," while others had the sense of physical contact with the breast. Almost all the women who had such a dream said they were convinced the dream was giving them an important warning.

One common dream narrative that has been a source of fascination and dread for thousands of years involves teeth, usually teeth falling out. Through the ages, dreams of losing your teeth have been interpreted as foreshadowing unpleasant events, such as a death in the family or the loss of property. A book from 1633 called *The Countryman's Counsellor* says a dream of bloody teeth predicts the dreamer's own death. The internet offers many, many more interpretations related to teeth dreams.

However, the real reason for these dreams could be much more mundane. Teeth dreams may be related to dental irritation during sleep. A study of 210 college students headed by a couple of Israeli researchers found dreams involving teeth were related to sensations of tension in the teeth, gums, or jaws upon awakening, which could be related to clenching or grinding teeth during sleep.[2] If other research confirms these findings, the cause of the iconic dream about losing your teeth could be very ordinary.

As for my patient, I never saw him again, but I could predict what his coming years looked like, because his dreams served as a roadmap. He would develop a progressive neurodegenerative disease that would damage his mind and ultimately lead to his death. As my life and career moved on, any time themes of mind–body connections came up in the scientific literature or just casual conversation, I would think about this patient—and how the brain's earliest warning flares of impending disease can be changes in dreams and dreaming.

Yet dreams have another, potentially more important connection to our health, through their ability to uncouple the mind from the body.

Dreams Help Us Cope with Emotional Distress

Almost all of us have dreamed at some point of showing up late for an exam, or being naked in public, or missing a plane or a bus. In our dreams, there is nothing stopping us from expressing our worst fears, or exposing our true emotions or our ugliest thoughts. In this way, they offer us a risk-free method of processing heartbreak, ill health, and other dire circumstances.

Consider dreams and divorce. No doubt, divorce is one of the major upheavals of adult life, a stressful event that upends the central relationship in a person's life and has a profound effect on our health. Large studies have found divorce, on average, has the same effect on life expectancy as obesity or excessive drinking. Some divorcees are able to get through a marital split well, while others fare quite poorly. The difference between those able to rebound and those who are not goes well beyond their state of mind. How might dreaming help us emerge from this potentially devastating life event in a healthy place?

One in-depth study of women going through a divorce found those who fared best in moving on with their lives may have continued dreaming about their former partners—but they no longer responded in a negative way to these dreams.[3] They had become emotionally neutral when the ex appeared in their dream. This emotional indifference was liberating and allowed these women to move past the divorce. Dreaming about an ex was not a sign of longing or regret. The emotion, rather than the content of the dream, was the key to understanding these thoughts.

Interestingly, researchers have found that dreamers coping

well with the end of a marriage are more likely to remember their dreams. It's possible dream recall enhances the potential therapeutic value by allowing the dreamers to contemplate them by day. These study participants were trying to cope with an emotional and unprecedented occurrence in their lives. Remembering the indifference they felt in their dreams about their former partner could be cathartic.

I could make the argument that talk therapy mimics dreaming, allowing for a safe environment to express ourselves, to work out different what-if scenarios, and to explore our emotions. Dream content can also be a valid topic for discussion in therapy, not because they reveal repressed desires, as Freud speculated, but because they expose true emotions. Clara Hill, an influential and now retired professor of psychology at the University of Maryland, argued that dreams are useful in helping people understand themselves more deeply, given that they are personal and can be "puzzling, terrifying, creative, and recurrent."[4] But Hill conceded therapists often feel unprepared to work with clients on their dreams because dreams are not covered in their training.

In addition to the potential therapeutic power of dreams themselves, our brain chemistry undergoes beneficial changes when we are experiencing our most emotional dreams. During REM sleep, the brain switches off the anxiety-triggering chemical adrenaline. There's no other time of the day when this happens. In this way, dreaming may serve as a sort of exposure therapy—dampening the emotional poignancy of dream experiences. As a result, people report fewer negative emotions after sleep and dreaming.

Sharing Your Dreams Creates Intimacy

Dreams offer an intimate view of your inner world, so sharing them can be a sign of trust, vulnerability, and emotional intimacy. Maybe this is why studies have shown sharing dreams with a partner can be an excellent way to enhance your relationship. Dreams also have the benefit of being symbolic, rather than literal, allowing feelings and family issues to be discussed openly but without blame, defensiveness, scapegoating, and power struggles. Emotional intimacy and satisfaction in a relationship go hand in hand, so it seems natural that sharing dreams would have a positive effect on a relationship. Few experiences are as private and potentially revealing.

One study compared couples who shared events of the day for thirty minutes, three times a week, with couples who shared their dreams for thirty minutes, three times a week.[5] Both groups made gains in marital intimacy and marital satisfaction, but the intimacy scores of the dream group were significantly higher. In a telling example, both partners in a couple married for more than ten years reported a need for more intimacy. The husband felt he had been completely open with his wife, but she did not feel emotionally bonded to him. When he shared his dreams, he revealed a different side of himself. By day, he was reserved and serious. In his dreams, he was effusive and rebellious. These dreams excited both partners and generated new life in their marriage.

A social worker at a maximum-security women's prison in the United States started a weekly dream group and found that sharing dreams generated a sense of trust, community, and connection among the inmates.[6] Being able to share dreams also helped them express emotions without fear of embarrassment, and cope with the hardships of being incarcerated. One inmate said the group helped her understand how events from her past led her to be imprisoned. Another inmate said, "I like that it is a safe place to share things. People are open-minded and they aren't judging you. They are supportive of you and your dreams."

Researchers at Swansea University in the U.K. studied the benefits of dream groups like this.[7] They found that sharing and discussing a dream can lead to important insights about one's waking life that may not have been gleaned without the benefit of a group. Dream groups also led to empathy toward the dreamer and a social connection between the person sharing the dream and the listeners.

Montague Ullman was a psychiatrist who started the Dream Laboratory at the Maimonides Medical Center in Brooklyn, New York, and promoted the benefits of dream sharing groups.[8] He developed a process for these groups; the same process can be used by partners in a relationship.

The first step is for the dreamer to recount their dream in full, without any interpretation. If there are characters in the dream, the dreamer says if they are real people or not, and, if so, what their relationship is to the dreamer.

Next, each member of the group asks themselves, "If this were my dream, I would feel . . ." and "If this were my dream, the symbols would remind me of . . ." They do this without looking at or talking directly to the dreamer. According to Ullman, this both indicates that members of the group are taking the dream seriously and, sometimes, can produce an observation the dreamer feels is insightful. The dreamer is then invited to respond.

Finally, the group can ask the dreamer questions and help them find connections between their dream and their waking life and the dream's potential meaning. Ullman suggests the most important skills for a dream group are being good listeners and asking questions that result in relevant information.

Dream groups offer an opportunity to build community and understanding around our most private act, dreaming. Sharing dreams allows others to understand us in a new way, just as understanding our own dreams is to know ourselves. To do both is, in my view, part of a life fully lived.

How Dreams Gauge Depression and Addiction

Depression affects how we see the world, our motivation, and, of course, our mood. By day, depression can flood us with feelings of despair, emptiness, and hopelessness. At night, this potentially overwhelming emotional burden can carry over to our dream lives.

Unsurprisingly, the imagery in the dreams of people who are depressed tends to be dark. Even people who are sad during their waking hours but not clinically depressed tend to have more negative emotions in their dreams. Likewise, people who report unpleasant moods when they are awake have more aggressive content, negative emotions, and misfortune in their dreams.

Dreams can also serve as a gauge of psychological wellbeing. Dreams can deliver an ominous warning to people with serious cases of depression. Someone suffering from a major depressive disorder experiences nightmares more than twice as often as someone who is not clinically depressed. The molecular mechanism for this is unclear.

More worrisome, though, is that nightmares seem to raise the risk of suicide or suicide attempts in people with depression.

One study looking at the dreams of individuals who were not depressed, depressed patients, and suicidal patients found dream narratives were a powerful indicator predicting suicidal behavior. The dreams of suicidal patients had a higher frequency of violence, gore, and murder. Among adolescents, frequent nightmares have been associated with subsequent suicide attempts and non-suicidal self-harm, which presents a crucial opportunity for early intervention.

Depression also changes how we sleep and dream in surprising ways. In people who are depressed, the architecture of sleep is altered. There is less time in deep sleep before the arrival of REM sleep, and an increase in the length and emotional intensity of REM sleep. Using non-invasive imaging techniques such as fMRI, which measures changes in blood flow associated with brain activity, researchers found differences between the emotional, limbic centers of the brain in non-depressed and depressed dreamers. The emotional, limbic centers of the brain were more active during REM sleep than during waking hours for all the dreamers. The emotional activity was highest among people with depression.

The REM dreams in depression typically become more negative with each ninety-minute sleep cycle, perhaps because these dreams focus on negative memories, becoming a negative circuit of anxiety and fear. This could be why depressed patients often struggle with mornings in particular, not from disrupted sleep but from the negative emotional milieu of their last dream before they rise.

Some people with depression report actually feeling better after going without sleep—at least for one night. Researcher Rosalind Cartwright decided to see if shortening REM sleep would help people who were clinically depressed.[9] Wake up most people in the middle of their vivid dreams, and they are tired and irritable when they get up the next day. Interrupt REM sleep in people who are clinically depressed, and their mood and energy levels are actually better in the morning. As you'll recall, in people who are not depressed, dreams can serve as a sort of nocturnal therapist, dampening negative emotions. This function of dreaming appears to fail in people with depression. Cartwright concluded that interrupting the emotional dreams of people with depression may have prevented them from reaching their negative endings.

But her findings do not provide a recipe to treat depression through sleep modification alone. In practice, it would be very difficult outside a sleep lab to shorten only REM sleep. Also, when we are deprived of REM sleep, the brain is starved of it. The moment we have the opportunity to sleep fully, we will make up for lost REM and the lost dreaming that accompanies it.

Dreams can also offer important tells about addiction. Drinking or drug dreams are common among addicts early in their recovery, especially those with more extensive histories of drug or alcohol problems. In fact, early in abstinence, drinking or drug dreams are typically more common than when the patient was actually consuming alcohol or drugs. It's as though their dreams are fulfilling cravings no longer satisfiable in their waking life. These dreams can be unnerving and produce intense feelings of fear, guilt, and remorse—until the recovering addict wakes up. The drinking and drug dreams usually taper off as the cravings subside.

You might assume that an addict dreaming of using again is a bad sign, but the opposite is true. Dreaming about drinking or drugs is considered a good prognostic sign for addicts in treatment. Feeling relief that the drug or alcohol use was only a dream signals a changed outlook. That's especially true if the recovering addict refuses to use drugs or alcohol in their dreams. A Brazilian recovering from a crack cocaine addiction described it this way: "I knew I could not use the drug in my dream. I took the drug in my hand, but I gave it to someone else. So it's cool that my subconscious is changing my way of thinking and acting. I even wake up happy to know that I did not use the rock in the dream."[10]

Dreams Can Warn Us of Pending Brain-related Illness

Although rarely discussed by clinicians with patients, dream dysfunction is also part of the final chapters in Parkinson's. While the most obvious signs of the disease are worsening physical symptoms, such as the loss of balance and coordination, the inability to walk without assistance, and a soft voice that trails off, nearly 80 percent of patients with Parkinson's dementia suffer from intense nightmares. And action-filled, aggressive nightmares can be the first sign of the late and incapacitating stage of the condition.

As we learned earlier, these patients also often tell of animals returning as the characters in their dreams, something usually experienced by children. And like the dreams of children, the animals in dreams are not their pets or domesticated animals but wild animals. The return of animal characters as the brain deteriorates makes me wonder if we revert to a more primitive version of ourselves, to the brains humans were living with not thirty thousand years ago but thirty thousand generations ago. Could it be that the developing brain and the ageing brain both dream of animals as a cognitive inheritance from our ancestors—from a time of rapid brain evolution when beasts and early humans coexisted? It's not such a wild question. Some dream disorders such as nightmares can cluster in families and are genetically heritable.

While the DEB disorder in middle-aged men will almost inevitably lead to Parkinson's disease, changes in dream patterns can also warn of worsening symptoms in another much more prevalent disease where mind and brain deteriorate: Alzheimer's.

Using exotic imaging techniques, we can now create a heat map of sorts correlating to the metabolic activity throughout the brain to measure energy consumption. The more energy being consumed, the more active that part of the brain. The heat maps show highly activated areas as red, while inactivated areas are blue. In patients with Alzheimer's, researchers found something striking. The areas of the brain that were blue on the heat map—those areas that were dormant—overlapped with the Imagination Network. Atrophied by Alzheimer's, the Imagination Network struggles to become activated. This may play out when Alzheimer's patients sleep.

But does Alzheimer's lead to dream loss, does dream loss lead to Alzheimer's, or does one play off the other in a downward spiral? Some scientists are now asking the question, could the lack of dreams exacerbate brain decay? Others have gone even further to suggest that Alzheimer's itself may be a disease of lost dreams: Along with memory loss, Alzheimer's also results in a loss of emotional regulation, something dreams assist with nightly. So it seems possible to me that dream decay affects emotional regulation in people with Alzheimer's. Because of the associated memory impairment, we may never be able to tease out whether these patients are dreaming less or remembering less or both, since the brain and mind are inseparable and reciprocal.

The link between our dreams and who we are is more

intertwined still in people with dissociative identity disorder, formerly called multiple personality disorder. Dissociative identity disorder is a mental illness in which an individual has separate and unique personalities that control their behavior at different times. It turns out that these personalities, called alters, often show up as characters in dreams before taking over the person's behavior during waking hours. These dream characters could be considered prototypes, a trial run for the alters who will emerge during the day.

Scans in patients exhibiting multiple personalities reveal intact and whole brains, which suggests that the alters are not the product of some sort of literal split or fissure in the brain. Interestingly, I've performed surgery where I've needed to "split" the left and right cerebral hemispheres, or even remove an entire hemisphere, and these patients did not report altered dreams, let alone new identities. The alters in dissociative identity disorder are something more intriguing than a physiological anomaly. They are a creation of the dreamer.

People with dissociative identity disorder can dream in other, different ways. Sometimes, one alter will show up in another alter's dream. Psychologist Deirdre Barrett has studied how different alters can recall the same dream from different perspectives.[11] For example, one patient described a dream in which she was a little girl huddled under a bed, worried someone would harm her. Another alter recalled the same dream, but she was a kid trying to distract the frightened girl, while a third was trying to menace the girl under the bed.

Schizophrenia is a serious condition that causes people to interpret reality abnormally, and it can also reveal itself in dreams. People with schizophrenia can hear voices or have the sense that someone is out to get them, and this troubled and distorted outlook of the world carries into their dreamscape.

Dream reports in schizophrenics are truly chilling. They can be full of aggression and sadism, often with images of mutilation. For most people, three-quarters of the characters in our dreams are people we recognize either personally or for their social role, such as a bank teller, a teacher, or a friend. For those suffering from schizophrenia, dreams are populated with a higher than usual number of strangers, often male and appearing in a pack. As patients with schizophrenia are treated with antipsychotic medications and their clinical condition improves, their dreams become less frightening and more emotionally positive, though they still report seeing more strangers in their dreams.

Given all the evidence that dreams can offer clues to our physical health, including the progression of Alzheimer's, dissociative identity disorder, and schizophrenia, it's puzzling to me why physicians don't ask patients about their dreams as a routine part of a medical evaluation.

When Dreams Harm Us

The occasional nightmare is common and can be brought on by stress or anxiety. By and large, they are harmless. They wake us up and are frightening but they are unlikely to affect our overall health or wellbeing. Nightmares that are part of a nightmare disorder are different. These are recurring and distressing nightmares that cause problems with how well we function during the day. They are distressing enough and frequent enough that some develop a fear of going to sleep. These nightmares are a cause for concern and should be addressed with your physician or a therapist. Otherwise, there's the risk they could cause a spiral of sleeplessness, daytime drowsiness, and anxiety.

Nightmares can be a measure of our emotional wellbeing. If we begin having frequent nightmares when we had few previously, that is something we need to pay attention to. If the pattern of nightmares we experience changes suddenly, that's another cause for concern. The nightmares may be a warning flag for something more serious going on with our mental health, such as depression. An estimated one-third of psychiatric patients experience frequent nightmares. Paying attention to nightmares, in my opinion, is no different than paying attention to headaches. If we go from having only the occasional headache to having frequent headaches, it's something your doctor should know about.

Almost three-quarters of people with PTSD experience frequent nightmares. Unlike the fever that accompanies an infection or the pain that comes with a physical injury, nightmares are not merely a symptom of PTSD. They can exact real emotional harm.

One of the hallmarks of PTSD is experiencing recurring dreams that replay the traumatic event again and again. The dreams are accompanied by fear, rage, or grief at night, and by hyper-arousability and anxiety during the day. PTSD nightmares are different from nightmares not provoked by trauma, which can be beneficial and even play a fundamental part in childhood development, as discussed in Chapter 2. Bessel van der Kolk, the psychiatrist who wrote *The Body Keeps the Score*, says trauma is not about the past but how the trauma lives inside you and that "dreams themselves could be traumatic for the dreamer."[12] In other words, dreaming of an event can re-traumatize the dreamer. The evidence for this is straightforward. During nightmares, our heart rate and breathing quicken as though we are experiencing the event itself. The parts of the brain activated during dreams match how the brain would react when we are awake. For example, running triggers the motor cortex. Experiencing fear triggers the amygdala.

On the other hand, dreams also have the power to recast traumatic events over time in a way that is therapeutic. Almost everyone experiences trauma in their lifetime, but the range of our responses in both our waking and dreaming lives is enormous. Some of us are able to rebound after a trauma, such as a car accident, the sudden loss of a loved one, or being the victim of a crime, in a psychological response called post-traumatic growth. Others don't fare as well. We can learn a lot about how well we're dealing with a trauma by paying attention to our dreams and determining whether they are symbolic or realistic.

Because the emotional focus of someone who has recently experienced trauma is clear, survivors of recent acute trauma are ideal subjects for dream research. American researcher Ernest Hartmann collected dream series lasting between two weeks and two years from forty such individuals.[13] What he found was that healing from trauma typically means moving from more literal dreams to ones that are represented visually in some other way. Dreams shift from replays or close facsimiles of an event to narratives that are symbolic.

One very common dream in the wake of trauma features a tidal wave. It has been reported by victims of all kinds of trauma. According to Hartmann, the dream narrative goes like this: "I was walking along a beach with a friend, I'm not sure who, when suddenly a huge wave, 30 feet tall, swept us away. I struggled and struggled in the water. I'm not sure whether I made it out. Then I woke up." Hartmann found trauma survivors also dream of being swept away in a whirlwind. Hartmann reported that the first four dreams experienced by one woman after being brutally assaulted were of being attacked by a gang, being choked by a curtain, being in the path of an oncoming train, and being caught up in a whirlwind. Disturbing as these were, they were a sign of healing.

Over time, as the trauma becomes less immediate and as the emotional impact of the event shifts, the images in our dreams shift as well. Dreams that begin contextualizing fear or terror may shift to embody helplessness and vulnerability, which might be represented in a dream of a small animal dying on the side of the road or walking in a large field with no shelter in a rainstorm. Then come dreams in which the central image represents survivor's guilt, and then grief.

It's remarkable that given the trauma most of us will endure at some point in our lives, we don't all get PTSD. The variables of who will get PTSD after exposure to a traumatic event and who will not remain elusive. That makes it hard, if not impossible, to predict who will be unable to shake traumatic memories—and the accompanying nightmares—and who will move on. But a recent breakthrough in neurobiology has identified a single molecule, neurotensin, that could serve as a kind of molecular switch.

Hao Li, a researcher at the Salk Institute for Biological Studies in Southern California, led a team that studied how positive and negative memories are encoded.[14] They concluded a signaling molecule called neurotensin acts like a switch and determines in the moment whether a memory will be encoded as negative or positive by the amygdala, the part of the brain responsible for the emotional imprinting of memories. The finding that one neurotransmitter can indelibly mark an experience may open the door to understanding the biological underpinning of PTSD. With PTSD, maybe neurotensin has overloaded the brain with negative signals. If that's true, neurotensin could offer a path to a new treatment. It is exciting to consider that by modulating neurotensin we may be able to treat the traumatic replay of nightmarish memories in PTSD.

Untamed, nightmares can lead people suffering from mental illness to a very dark place. They have the potential to escape the world of dreams and enter into waking life as psychotic episodes. In one case report, a 78-year-old man admitted to the hospital after a suicide attempt had been suffering from nightmares for three years. The nightmare was always the same: A man wielding an axe and accompanied by big dogs was chasing him. The dream was so terrifying, the patient tried to avoid sleeping altogether. In the two weeks before he was admitted to the hospital, he had repeatedly awakened to auditory and visual hallucinations involving the man and his dogs. Finally, he attempted to kill himself with an axe "to complete the work for this man." There have been other cases reported where nightmares transition into psychotic episodes, underscoring not only how fluid our dreaming and waking lives can be, but the surreal and piercing nature of nightmares.

Dreams Can Make Us Whole

In dreams, we can make ourselves whole in ways that might be expected—and ways that seem impossible. In their dream bodies, amputees talk of once again having all their limbs. The missing arms and legs have been restored. Even though the sleeping brain is receiving no signals from the amputated limbs, the dreaming brain is able to use these non-existent limbs as though they had never been lost.

In various studies, amputees report dreams that would be impossible in the waking world. A man with an amputated arm dreamed of smashing a mosquito with both hands.

Another, of shifting gears as he drove a Ferrari Testarossa and, later, of pouring a drink for a friend, holding a champagne bottle in his right hand and a glass in his left hand. A woman with most of one leg amputated dreamed she ran from a plane after it flew too low overhead.

When dreams make us whole, what occurs in the dreamscape is nothing short of magic. Two women confined to wheelchairs as the result of chronic spinal cord injuries independently reported something astonishing. They each told of having dreams that included their wheelchairs. However, they were rarely seated in them. In their dreams, they preferred pushing the empty wheelchair.

For patients with Parkinson's and dream enactment behavior, dreams enable them to overcome the limitations of their waking bodies. In defiance of all scientific logic, they exhibit something called paradoxical kinesis. By day, their limbs may be rigid and tight, their movements slow, almost ossified. This isn't for lack of will. The signaling from the brain to the body is breaking down. When these Parkinson's patients dream, and they act out their dreams as a result of dream enactment behavior, their movements are not slow or jerky, as you might expect. They can move quickly and fluidly. The tremors, weakness, and rigidity they experience during the day are gone. Their voices, too, are transformed. Quiet and tremulous by day, they are now loud and clear as they shout in their sleep. How this paradoxical kinesis is possible remains an enigma.

As we learn more about the neuroscience of dreaming, we are also learning about the potential inside our bodies and minds that only dreams reveal and release. It's not just our imagination, narratives, and relationships that are boundless in our dreamscape. The dreaming brain holds other powers.

Since meeting that patient nearly twenty-five years ago, I have immersed my life and study into care and investigation of the human brain—the human person—from a breadth of scientific perspectives. The more I have learned, the more I stand in awe and wonder at the mystery of our minds.

One such capacity, the ability to awaken inside our dreams, and control the direction of dream events, feels more like magic than science. Despite its description for millennia, only in this decade are we able to scientifically investigate and prove that our brains can indeed be dreaming—and also partially awake.

6.

Lucid Dreams: A Hybrid of the Waking and Dreaming Minds

In 1975, an experiment rocked the field of neuroscience.[1] The goal was nothing less than to revolutionize our understanding of waking, sleeping, and dreaming: to show that dreamers could become self-aware while they were still dreaming, and to prove it by communicating with the outside world. In other words, to prove that lucid dreaming was real.

A participant named Alan Worsley had fallen asleep in a sleep lab in England with very specific instructions.[2] He was asked to move his eyes back and forth when he became aware he was dreaming—while still inside the dream. To show his eye movements weren't random, he was instructed to move his eyes smoothly left-right-left-right in a way that he had rehearsed while awake. These deliberate eye movements would be impossible to confuse for the erratic eye movements of REM sleep.

The commands focused on the eyes because muscles are paralyzed during REM sleep—except those that control eye movement and breathing. This makes dreamers similar to those rare cases of locked-in syndrome, where a catastrophic central brain injury leaves people paralyzed from the eyes down and only able to communicate with blinks or eye movements. It was an audacious attempt, but researcher, Keith Hearne, knew big claims needed big proof. Even with the eye movements, the appropriately skeptical scientist should ask: How could you know he hadn't awoken just long enough to move his eyes left-right-left-right?

It's a reasonable question, one Hearne was anticipating. Because the lucid dreamer's scalp was wired with dozens of electrodes, Hearne was able to record the electrical signature of sleep throughout the experiment, registering spikes of electrical activity called sleep spindles that can't be faked. A set of electrodes tracking electrical activity in the participant's muscles showed atonia, the near total paralysis of the body. Another measure of electrical activity, or in this case a lack of electrical activity, that also can't be faked.

What is lucid dreaming? Lucid dreaming is the experience of dreaming and *knowing* you are in a dream. To lucid dream is to enter a paradox that seems more mystical than real, a dual consciousness that straddles the vivid, illogical dreamscape and the insight that you—the dreamer—are both the creator and an actor in this imagined world. In some cases, lucid dreamers are able to take lucid dreaming a step farther and control the action within the dream, a type of real-time dream navigation.

Lucid dreaming is not some new age adventure discovered by hippies or gurus. It has been around since antiquity. Long before Hearne and modern science arrived on the lucid dreaming scene, the phenomenon was well known. Aristotle referred to lucid dreaming in his fourth-century BCE treatise, *On Dreams*. He wrote, "Often when one is asleep, there is something in consciousness which declares that what then presents itself is but a dream."

Despite centuries of accounts of lucid dreaming, the neuroscience community mostly viewed claims of lucid dreaming with skepticism, as did I. By definition, dreaming occurs outside of our conscious awareness. Maybe people who think they are lucid dreaming are simply dreaming they are lucid, like a dream within a dream. Or maybe they woke up briefly then went back to sleep and were under the misimpression that they had been conscious within the dream. Maybe they were not yet asleep, or in the process of waking up, and what they thought was a lucid dream was more like a half-awake vision.

The other problem researchers faced was that even if lucid dreaming was possible, how could you prove it? After all, how could you objectively demonstrate someone was lucid dreaming without waking that person up? And once you woke them up, you were relying on their subjective recollections. As Hearne was aware, some test subjects are eager to please. They might simply report having had a lucid dream because that is what the researcher wanted to hear. Just as vexing: If the dreamer remained asleep and the body is paralyzed during REM sleep, how could they possibly signal that they had entered a lucid dream?

For years, researchers had tried various methods that would allow lucid dreamers to communicate. One tried to have dreamers raise a finger. Others tried to train people to make other small movements, or trigger a micro-switch taped to sleepers' hands. None of these methods worked. The paralysis of REM sleep can't be overcome with training or willpower, and movements like these are simply impossible. Remember, the body during REM is very much like someone who is paralyzed below the eyes. It was researcher Keith Hearne—only a graduate student at the time—who realized eye movements might be the key.

Hearne was a novice researcher, an unlikely candidate to usher in a new realm of dream neuroscience. He met his test subject, Worsley, by chance. The 37-year-old helped Hearne and his wife move their belongings to a new house and mentioned that he had lucid dreams. When Hearne started doing research on the subject, Worsley volunteered.

At the time, eye movements were already being measured in sleep labs to show when REM sleep began, using a device called an electrooculogram. The device works by placing electrodes on the skin near the edge of each eye. When the eyes move, even under closed eyelids, there is a change in the electrical signal. The results are recorded on a computer, or as was the case when Hearne conducted his experiment, as lines scratched out on a roll of paper.

Normally, during REM sleep, eye movements are haphazard. There is no pattern on the electrooculogram chart. Therefore, Worsley was asked to move his eyes back and forth. These deliberate eye movements would be impossible to confuse for the incidental eye movements of REM sleep, and would jump out on the electrooculogram chart from the indistinct squiggles produced during normal REM sleep.

The first night Hearne tested his idea that lucid dreamers could signal using eye movement, the night passed without a single signal from Worsley. A little after 8 a.m., Hearne figured the experiment had failed. He was folding up the chart paper when suddenly Worsley had a lucid dream and attempted to signal. But it was too late. The equipment had already been switched off.

A week later, Hearne made another attempt. Again, Worsley had a lucid dream a little after eight in the morning, but this time Hearne was ready. The eye movements produced large distinct zigzags on a scroll of paper on the electrooculogram. Hearne watched with astonishment. Moments earlier, he had been half asleep watching the ink tracing scroll past, as he'd been up all night monitoring the recording equipment. What he saw on the chart paper jolted him awake, knowing he was witnessing history. He later wrote he was as excited as if the signals had come from another solar system.[3] That moment officially opened the door to the rigorous scientific exploration of lucid dreaming.

The slashing strokes up and down on the scrolling chart paper would become seismic shocks in the neuroscience community. It was the first time someone had signaled in real time while dreaming—proving that in at least one case, in one person, it was possible to be awake inside a dream.

More than two and a half thousand years after Aristotle wrote about lucid dreaming, Hearne published his findings. His work was peer reviewed, contested, and accepted reluctantly after other researchers validated and expanded on his work with lucid dreaming tests of their own, using the same left-right-left-right eye signals Hearne introduced. These have become the gold standard for lucid dream research, a kind of Morse code now used in sleep experiments around the world. Left-right-left-right in a sleep lab means: I am lucid dreaming.

How Lucid Dreams Emerge

Since then, the scientific understanding of lucid dreaming has expanded and grown more sophisticated, resulting in a massive field in science today. In the four decades since Hearne's experiment, we have learned a lot about the subject, yet much remains unknown. As researchers dig deeper into the mysteries of lucid dreaming, trying out different imaging techniques and giving lucid dreamers new challenges, they are also hoping to learn more about how the brain itself functions. It's as though lucid dreaming has offered a window into the workings of the brain we couldn't see previously, and to which we didn't have access.

Almost everyone says they have experienced spontaneous lucid dreaming at least once in their lifetimes, and about one in five people say they have at least one lucid dream per month. Lucid dreaming appears more commonly in women than men, is more common in children and tends to taper off after adolescence. With lucid dreaming, it is as though consciousness has found a new dimension: A liminal, hybrid state where we can be dreaming but awake, or awake in a reverie.

But how is lucid dreaming possible at all? How can a dreamer become aware that they are dreaming while technically still asleep? And when they do, why doesn't this awareness break the spell and awaken the dreamer? What is happening in the brain that lets the mind be both partially awake and asleep?

As we know, during ordinary dreams, the Imagination Network is activated and the Executive Network is shut down. Since the dorsolateral prefrontal cortex—the rational, reasoning, skeptical part of our brain—is inactive, we are not troubled by the unreality of dream narratives, and at a more fundamental level, we also lack awareness we're dreaming, thereby fully inhabiting the dream experience. In lucid dreaming, however, something happens that breaks down this suspension of disbelief. Lucid dreamers will often report a moment when the scene in their dream was so unrealistic, they realized they must be dreaming. Common experiences in dreams that can trigger lucid dreaming, or "dream signs," include strange emotions, impossible actions, odd or shape-shifting bodies, or bizarre settings and situations. But what's interesting is that these strange occurrences are more or less normal in dreams.

What then is happening in the brain that triggers this leap of understanding, this momentary clarity that what the dreamer is experiencing is only a dream? If dreams are bizarre in general, what is the difference in bizarreness that qualifies it as a "dream sign?"

We don't know the answers to these questions, but researchers have found several clues that offer us glimpses into how lucid dreams differentiate from normal dreams. Brain imaging, for instance, suggests that the Executive Network may be partially switched back on during lucid dreaming. Most of what we know about the science of lucid dreaming comes from electrical signals recorded on the scalp by an EEG. One difference in these EEG recordings of lucid dreaming, compared to ordinary dreaming, is an intensification of the higher-frequency brainwaves in parts of the prefrontal cortex. As we learned in Chapter 1, this area houses the logical part of the brain, which we normally deactivate when dreaming.

Researchers have also taken a step closer to understanding what can initiate a lucid dream. Using transcranial stimulation, a non-invasive procedure that sends a faint electric pulse into the brain from outside the skull to activate various parts of the prefrontal cortex, scientists have found that electricity increases lucidity, even in dreamers without lucid dreaming experience. Transcranial stimulation is a developing technology to treat conditions such as depression and migraines but has also shed light on the workings of the brain and mind, so it's not unreasonable to think we may someday be able to use a device to generate lucid dreaming at will.

For now, lucid dreaming remains the province of a relatively small number of people able to regularly access this dual world. Lucid dreams are an amazing mental feat, but they seem to be fragile, and they are not always fully accessible. In one clever experiment,[4] lucid dreamers were asked to evaluate a scene when they were awake, such as a room at home, and really soak in the details. When they were dreaming lucidly, they were asked to change their dream environment to resemble this scene from memory. These dream "reinstatements" were typically inaccurate, and even if the lucid dreamer was aware they were flawed, they remained so. This is how one test subject described his lucid dream:

> I opened the door and the room was empty . . . I closed the door and tried to make things appear as they were in the room . . . I would close my eyes, think of an object that I could remember and open my eyes and it would appear. First it was the wooden desk with the fruit . . . I kept closing my eyes and trying to make it perfect, but then things got out of control.

This lucid dreamer never did manage to recreate the room correctly in his lucid dream, and others experienced similar difficulties.

Despite the sense of awareness that comes with lucid dreaming, your body still behaves as though what you're experiencing is real, as it does with normal dreaming. For instance, when lucid dreamers hold their breath in a dream, their bodies show a central apnea. When they exercise, their heart rate goes up. The breathing of lucid dreamers quickens as they dream of sex. Lucidity appears to enter into a normal dreamscape, yet somehow the awareness of being in a dream does not diminish the body's response to the dream narrative. This is what gives lucid dreamers both the insight of being in a dream and the full-body, visceral response of being there.

In thinking about this, a question that comes to mind is whether the dreamscape feels different to the lucid dreamer once some awareness has returned. An elegant answer was offered by researchers who didn't use any exotic imaging or other cutting-edge techniques in their studies of lucid dreaming.

Instead, they took a simple but essential monitor used in sleep studies, the oculogram, which tracks eye movements.

If you see a flock of birds flying slowly in the distance when you're awake, your eyes will follow the flight smoothly. If you *imagine* the same birds flying across your field of view when you're awake, your eyes will not track smoothly. There will be skittish jumps, called saccades. However, when lucid dreamers imagine the flock of birds flying across their field of view, they track the birds smoothly. This can be seen in their eye movements, which show they are fully immersed in the dream world and tracking movement as they would with real birds. If the awareness that they were lucid dreaming somehow made the experience less real and more like an act of imagination, the eye movement would result in saccades.

Even with all the research, it's unclear why we have lucid dreams. One theory suggests that lucid dreaming represents a true hybrid state, where the awareness of the waking brain is injected into REM sleep, possibly as the result of a particular brainwave pattern returning in the frontal lobes. Yet another tries to place lucid dreaming in a continuum of consciousness that includes dreaming, mind wandering, and waking. At least for now, these theories are all simply ways to conceptualize a strange and, for some, wonderful addition to their dream life.

Using Lucid Dreaming to Our Advantage

Throughout history, lucid dreaming has been a means of increasing spirituality, seen by religions as nothing less than a portal to enlightenment and the divine. In Buddhism, the Tibetan spiritual practice of dream yoga seeks to use lucid dreaming to gain spiritual insight. In fact, dreaming is considered to have more potential than waking for spiritual understanding. The 1,200-year-old teachings allude to lucid dreams as "the method that brings realization of great bliss," and advise adherents to "know dreams as dreams and constantly meditate on their profound significance."[5]

Amerindian people, Australian Aboriginal peoples, and Christian monks have also prized the ability to control lucid dreams as a vital aspect of their spiritual path. In this state, they are able to seek out ancestors, spiritual beings, or the divine.

In an interesting experiment, lucid dreamers were told to formulate a phrase they would ask in their lucid dreams to seek out the divine, such as: "I would like to see how the universe is run," and "I wish to experience the divine." They repeated the phrase during the day before their lucid dream. This facilitated lucid dreams where some dreamers reported experiencing the divine. Interestingly, the divine they experienced in these dreams corresponded to their waking beliefs. Those who believed the divine was a being tended to dream of the divine presence as one, while the remaining lucid dreamers experienced the divine in other ways. One lucid dreamer reported seeing the divine "as a moving picture with numerous interwoven cycles—like the workings of a clock. It is also like patterns of pulsating light and shadows moving in cycles."

Even when experiments like these do not produce profoundly moving experiences, they can still result in a deeper sense of wellbeing that is long-lasting. More than spirituality, surveys show an overwhelming majority of lucid dreamers believe the ability to dream lucidly is empowering, and they wake up in a better mood after a lucid dream. Lucid dreamers also say their dreams have contributed to their mental health, allowing them to make beneficial changes in their lives, and they credit lucid dreaming with inspiring them to take the chances necessary for these changes.

Given the impressions of lucid dreamers themselves, could lucid dreaming be used as a therapeutic tool? Could the ability to partially steer a dream be used to alter the emotional milieu of the dream itself? Could nightmare rescripting be possible not by autosuggestion, but within dream direction?

In Imagery Rehearsal Therapy, recurrent nightmares sufferers can literally rewrite those nightmares during the day to change the plot and their role in it while dreaming. In the same way, if you could become lucid during a nightmare, could you alter the plot and break its spell? That's exactly what Alan Worsley describes learning to do when he was only a five-year-old boy. When he had a nightmare, he would become lucid and yell, "Mother!" as a way of waking himself up. Therapists have trained people with chronic nightmares how to dream lucidly and found the practice helps. The benefit may extend beyond the nightmares to the accompanying anxiety and depression.

In fact, German researcher Ursula Voss found that teaching people with PTSD how to lucid dream helped them alleviate their symptoms in a number of ways.[6] As we learned, one of the hallmarks of PTSD is recurring nightmares revisiting the traumatic event, which has a secondary effect of making people with PTSD afraid of falling asleep. Empowering people with PTSD to control their sleeping thoughts through lucid dreaming allows them to change or end a recurring nightmare as it is happening. Instead of being victims in their dreams, they might call the police or disarm their assailant. Voss tells the story of a woman who made the person traumatizing her float in her dream, proving to herself it was not real.

The power to lucid dream gives PTSD sufferers the confidence to stop fearing sleep, and can make them more optimistic that they will someday be able to cope with their trauma.

Lucid dreaming could also have clinical applications. Researchers have reported, for example, that lucid dreams could help people with anxiety confront their fears or phobias, such as driving or heights or spiders.[7] They could "practice" driving, standing on a high ledge, or allowing friendly spiders to crawl on them in a safe environment, knowing it was only a dream.

Since the parts of the brain activated in a dream are the same ones that would be activated if we were really engaged in the behavior, it is possible that lucid dreaming could benefit people who have either had a stroke or suffered a serious injury. Could lucid dreaming be a new, pain-free venue for rehab? Opening and closing a fist in a lucid dream produces the same activation in the sensorimotor cortex as actually opening and closing when awake. If you were trying to come back from a sports injury, wouldn't practicing in a lucid dream be beneficial?

People who are paralyzed or otherwise disabled, too, might benefit from the power to move freely and at will, even if it is only in a dream. How liberating would it be for someone whose mobility is non-existent or severely limited to take control of their dream and run or jump? Similarly, lucid dreams could be used to help people who are in a partially comatose state, or people with locked-in syndrome, to move outside themselves.

The potential of lucid dreaming is not limited to therapeutic applications. It can also be used to improve performance. Many athletes use mental visualization while they're awake, using their imagination to simulate different scenarios. Lucid dreams can serve as another venue for neuro-simulation. Athletes could use lucid dreams to practice aspects of their sports that are potentially dangerous, such as a particularly difficult or challenging gymnastic routine.

Surveys of athletes who use lucid dreaming to practice specific skills in their dreams found that most believed it helped them improve significantly in real life, and some said it boosted their confidence.[8] One martial artist reported lucid dreaming helped him master a complex kick combination. As a bonus, he could work on it in his dreams without any risk of injuring himself. Other athletes in the survey took advantage of the dream environment to do things that they couldn't in the real world, like impossible mountain bike descents or alpine ski jumps.

Melanie Schädlich at the University of Heidelberg in Germany decided to test whether lucid dreaming practice could improve physical performance.[9] She asked lucid dreamers to "throw" darts and "toss" coins into cups that were placed farther and farther away in their dreams. What she found was the lucid dream practice really did help: Those who practiced in their dreams showed improvement in real life, as long as they didn't get distracted in the lucid dream. Although this was a relatively small study, lucid dreaming could become the next frontier for athletic training. Not only can athletes work on difficult skills without fear of injury, but injured athletes can "practice" before they are ready to return to their arena.

Schädlich and Daniel Erlacher conducted another study, this time seeking out musicians who were lucid dreamers.[10] But they found they didn't use their dreams to practice. Instead, the musicians said they were interested in playing in their lucid dreams for enjoyment and inspiration, rather than for improving their skills. In interviews with five musicians, they found the lucid dreams resulted in positive emotions and boosted confidence. Two of the five said they particularly liked improvising solos in their lucid dreams.

Because of the potential control lucid dreaming offers, this unique state of consciousness also creates enormous possibilities to foster creativity, even beyond normal dreaming. To take full advantage of the creative potential of lucid dreaming, you can ask yourself a question before bed as you would when you are priming a normal dream, only now you can potentially take control. Lucid dreams have the added advantage of being more memorable than a typical dream. In one case study, a computer programmer reported using lucid dreams to help design his programs. In his lucid dreams, he reportedly discussed what he was trying to do with Albert Einstein, and together they drew flowcharts on a blackboard until they found a solution.[11]

With that case study as their springboard, researchers at Liverpool John Moores University decided to see how well nine lucid dreamers did versus nine non-lucid dreamers in solving a task in their sleep.[12] For ten consecutive days, each night at 9 p.m. they received an email with their task. They were either asked to solve a logical puzzle or create a metaphor. For example, they might be asked to find the missing letter in a sequence or create a metaphor for such phrases as "a banknote floating on a river" or "a lighthouse in the desert."

The lucid dreamers were encouraged to believe there would be someone in their dream who "knows answers to many questions and is willing to help," maybe an older wise person, or a trustworthy guide. They were asked to find this figure. If they couldn't, they were told to go forward, turn left, find a door, go through it, and turn right. These elaborate instructions were designed to raise expectations among the lucid dreamers that they would find their guide. When they did, they were encouraged to ask this dream character to solve the problem they had been given. No matter what answer they were given, the lucid dreamers were told to thank the guide, wake themselves up, and write down the answer.

When the results were in, it seems the dream guides were not great at solving puzzles. Out of eleven answers they gave during the study, only one was correct. Because the Executive Network is only partially activated in lucid dreaming, it's quite possible the puzzles were simply too difficult for the lucid dreamers, guide or not.

Perhaps these imaginary guides would have fared better if the creative challenge had been less of a word problem and more of a visual one. For instance, Worsley, Hearne's test subject, performed lucid dreaming experiments that tended to involve challenging new ways to manipulate the visual environment of the dreams. In one, he would find a television in his lucid dream, turn it on, change the channel, and manipulate things like the volume, the intensity of the color, or the image on the screen. Worsley also says he has played the piano, walked through walls, created a flame by flicking his fingers like a lighter, and put his arm through a car windshield in his lucid dreams. He even passed one forearm through the other, and has extended body parts like his nose and tongue by gently pulling them out.

British artist Dave Green draws portraits of people in his lucid dreams, and he recreates the portraits as soon as he wakes up. Though he is a practiced lucid dreamer, Green says there are challenges to creating art in his dreamscape: Everything in the dream is in a state of flux and could transform into something else at any moment. He describes the process as "an interaction between my conscious and unconscious mind playing itself out on the page in real time."[13]

Worsley, too, has said the lucid state is tenuous, even for someone adept at becoming lucid in a dream. He says his level of lucidity can change from moment to moment. In that way, for Worsley anyway, the same dream lasting just a few minutes can be both lucid and not lucid.

The New Frontier of Lucid Dreaming

Aside from the eye-movement signaling when a lucid dream begins, researchers have no objective signs of what is happening inside a lucid dream. And it's not possible to signal when a lucid dream ends. It seems the fragility of a lucid dream likely rests on its very nature as a hybrid and delicate state of consciousness.

Even with limits such as the ephemeral qualities of lucid dreams, researchers have found new and inventive ways to take lucid dreaming farther than anyone thought possible. They've been able to train subjects, often students with no prior experience with lucid dreaming, to respond to flashing lights as they sleep with the left-right-left-right (L-R-L-R) eye movements. Some test subjects can even use the eye movements as a "timestamp" when they are beginning or ending prearranged tasks. That in itself is remarkable.

What is utterly astonishing is that researchers and dreamers have now engaged in two-way communication, back-and-forth interactions with prompts from researchers and responses from their dreaming subjects. This would have been considered an impossibility only a few years ago. Dreamers are able to process words or signals from the waking world while demonstrably remaining in REM sleep.

With their bodies paralyzed by REM sleep, dreamers have even answered spoken yes-no questions from researchers. In one study, the lucid dreamer used eye movements to respond to the question: "Do you speak Spanish?" The test subject later reported that he was dreaming he was at a house party and the question seemed to be coming from outside, like the narrator of a movie.

We're not yet sure how this is possible, but there are reports in the academic literature that offer some insight into potential neurobiological underpinnings. In one case study, a 26-year-old woman and 37-year-old man both had strokes in the thalamus. Following their strokes, they began having frequent lucid dreams. The lucid dreams lasted for about a month for each of these patients before tapering off, possibly as their brains healed. Could the lucid dreams in these two patients have been caused by a malfunctioning of the brain's built-in arousal mechanism?

Remember, when we're sleeping, we are not completely shut off from the world around us. Instead, a process called thalamic gating allows our bodies to monitor sounds for something alarming, or an unusual sound that signals danger. When noises or some other sensory information is deemed a sign of danger, the thalamus relays the information to the frontal lobes, arousing the sleeper.

Perhaps something similar is happening in the thalamus of healthy people who are lucid dreamers. Maybe lights, sounds, and voices that are normally filtered out during dreaming are now seen or heard, even if they are still incorporated into the dream setting. It's likely why people who are dreaming lucidly can hear questions from researchers as though they are coming through walls or in other unreal ways.

In an experiment at Northwestern University's Cognitive Neuroscience Program, PhD candidate Karen Konkoly was able to get lucid dreamers to do something mind-blowing: solve simple math problems while they were dreaming.[14] Dreamers were told ahead of time that they would be doing math problems in their dreams and were taught how to signal their answers. Moving their eyes left-right once equaled the number one. Moving their eyes left-right twice was two, and so on.

One lucid dreamer was given the prompt 2 + 1. She said she was looking at a house in her dream at the time. She incorporated the question into the number plates above the front door and signaled three by moving her eyes back and forth three times.

Because dreams don't have the same logic as our waking lives, even lucid dreamers don't question where the voice asking the question is coming from. They might hear it coming from the ceiling or through a car radio. One test subject conveniently dreamed he was in his math class.

However, the two-way communication from researcher to lucid dreamer is far from perfect. Out of thirty-one math problems, Konkoly's team only received six correct responses; they also received one incorrect response and five ambiguous responses. Most of the time, the lucid dreamer didn't respond at all. Still, this is a level of communication that had never been achieved before and was not even thought possible until recently.

Now, you may be wondering how lucid dreamers could do any math at all in the experiment. As you'll recall, calculation is something the dreaming mind can't do. That test subjects were able to do math during lucid dreaming is strong evidence that during lucid dreaming, the Executive Network is powered up just enough to allow for simple computation. This may also provide just enough self-awareness and critical thinking for the dreamer to become aware they are in a dream. These findings are astonishing, and perhaps they can point to the only possible conclusion: Lucid dreaming represents a distinct form of cognition, a true hybrid of the waking and dreaming minds.

If people in lucid dreams can successfully answer math problems, what else can they do? Could we one day be able to hear what people are saying in their lucid dreams? Though it sounds implausible, this, too, may be on the horizon.

One research team decided to see if lucid dreamers could say "I love you," in a lucid dream in a way that could be objectively measured.[15] Based on previous research, this should be impossible. Even if someone was able to speak these words in a lucid dream, how could lucid dreamers do more than signal L-R-L-R that the task was complete? How could the words themselves be measured?

In an attempt to decipher what was happening when the lucid dreamers slept, the researchers first recorded the fine facial movements around the eyes that accompanied each of them saying "I love you" when they were awake. These are among the few muscles that are not paralyzed during dreaming. These measurements when the test subjects were awake served as a physiological signature of sorts. Once the researchers had these, they recorded the muscle movements around the lucid dreamers' eyes as they slept. All four volunteers were able to say "I love you" in lucid dreams, and the words they uttered in the dream world were recorded in minute muscle movements around their eyes.

These dreamers showed lucid dreams weren't limited to responding to prompts from researchers. They could, potentially, initiate their own communication. They were for the first time communicating from the dream world to the waking world with spoken language, perhaps ushering in a new frontier for neuroscience.

The scientific community has come a long way on lucid dreaming in a short time. Long disdained by researchers as the province of mystics and cranks, they now embrace lucid dreaming as a novel form of consciousness worthy of serious research. As skepticism has given way to excitement, clever experiments are finding new ways to interact with the dreaming mind—revealing new aspects of dreams and dreaming in the process. Moreover, lucid dreaming is not something that can only be achieved in a sleep lab. It is within reach for all of us.

7.

How to Induce Lucid Dreams

Léon d'Hervey de Saint-Denys started recording his lucid dreams at age thirteen and continued until he had filled twenty-two volumes with elaborate dream reports covering 1,946 different nights. At first, his dream recall was sporadic. But the more he wrote down his dreams, the more he remembered them. By the 179th night, he was remembering them most nights. Not long after that, Saint-Denys had his first lucid dream.

At the time—the mid-nineteenth century—lucid dreaming was not thought possible. Even the term lucid dream would not exist for another half century. But six months later, Saint-Denys was lucid dreaming two out of five nights. A year later, he was experiencing lucid dreams three out of four nights.

In addition to becoming a frequent lucid dreamer, Saint-Denys learned to control his lucid dreams and used his experiences to test his theory that dreams were not the product of a supernatural or external force but the dreamer's own memories. He would pause his lucid dreams to study the surroundings and later compare them to his daily life. Saint-Denys also wanted to see if it was possible to do something in a lucid dream that he had never experienced in waking life. Toward that end, he jumped from a window, used a conjured sword to fight off masked attackers, and cut his own throat with a razor.

In 1867, Saint-Denys decided to share, anonymously, what

he'd learned from his intense study of sleep and dreams. He wrote a guide to lucid dreaming, *Les Rêves et les Moyens de les Diriger: Observations Pratiques,* or *Dreams and the Ways to Guide Them: Practical Observations*.

A century ago, a British woman named Mary Arnold-Forster followed in the footsteps of Saint-Denys. In her book *Studies in Dreams*, she describes how she used autosuggestion to help spark lucid dreaming. At bedtime, she told herself to notice her subsequent dreams. She became an accomplished lucid dreamer and particularly enjoyed flying, which she achieved by giving a slight push or spring with her feet.

Only one in five adults reports having even a single lucid dream in any given month, and the percentage of people who are frequent lucid dreamers and have several lucid dreams a week is very small, likely in the single digits. But the ability to dream lucidly appears to be inducible, a cognitive capacity that can be pursued, trained and brought to the fore with intention.

It also appears that our lifestyle and hobbies can influence how frequently we experience a naturally occurring lucid dream. For example, people who play video games have more lucid dreams than non-gamers. Perhaps they are more frequent lucid dreamers because in both lucid dreams and video games, the participant is controlling a simulated reality. Gamers sense of spatial awareness may also be heightened, which could help produce lucid dreaming. Athletes also tend to have a highly developed sense of spatial awareness, and they, too, are more likely to have lucid dreams. A study of professional athletes in Germany found they were twice as likely as others to experience lucid dreams.[1] Even more impressive, most of them did not make any special effort to become lucid. It just happened for them.

In my practice, certain medications in patients with cognitive decline, brain injury, or in the recovery phase after brain surgery have led to reported increases in both dreaming and lucid dreaming, especially those that modulate the neurotransmitter acetylcholine. We'll get into the neurochemistry later, but first let's look at the ways one can induce lucid dreaming without drugs.

How to Tell if You're Lucid Dreaming

Like Saint-Denys, researchers who want to study lucid dreamers have spent a good deal of time figuring out ways to improve the odds that one of their test subjects will become lucid on any given night. They have a professional interest: Any night a research participant spends in the lab and does not experience a lucid dream is a waste of time and resources. With that incentive, they've come up with several methods for inducing lucid dreams that require nothing more than your mind, and possibly an alarm clock.

The methods focus on the two essential aspects of this rare hybrid state. The first is that the dreamer should be in REM sleep because that's when lucid dreams typically occur. Several techniques to induce lucid dreams try to increase the chance REM occurs as close to the waking state as possible. The second essential aspect of lucid dream training is to achieve the insight that what is being experienced is a dream.

Let's walk through some of the methods researchers use to induce lucid dreaming. The simplest is called Reality Testing, and it rests on the fundamental aspect of lucid dreaming: the ability to discern the difference between waking and dreaming states. This insight that we're dreaming triggers lucidity. For instance, lucid dreamers will tell you they knew it was a dream when they saw a relative who had died long ago or they realized they were in a house that no longer exists or were in the middle of some other impossible scene.

Reality Testing attempts to heighten our awareness about the sleep and wake states by asking ourselves throughout the day, "Am I awake or am I dreaming?"

But if you ask yourself if you're dreaming and the answer appears to be yes, how can you be sure? Maybe you're only experiencing a dream within a dream. Or maybe you've been aroused from sleep and are in that fuzzy mental space between waking and sleeping. In the movie *Inception*, objects called totems were used to differentiate between reality and dreams. In our reality, we don't have totems like the movie, but lucid dreamers have figured out their own totems to reveal if they are in a dream. It turns out how we recreate reality in our dreams has some common and telling flaws.

If you think you're in a lucid dream, focus on your hands. For some reason, hands look strange in dreams. Count the fingers—there may be too many, or too few, or the number of fingers may change. Lucid dreamers report counting and recounting the number of fingers and getting different numbers each time, or fingers appearing rubbery as though they had no bones, or that they had fingers growing out of fingers. This strange phenomenon has been reported by lucid dreamers around the world and across cultures.

Could it be that hands simply take up too much mental processing power? After all, hands are exquisitely complex anatomy. Fingers can move independently, and we grasp in very particular ways. Hands are also mirror images of one another. This sort of left–right mirroring is common in nature, but to visually reproduce two hands accurately is not easy (just ask someone taking a drawing class).

Dreams attempt to reproduce reality from memory without the benefit of something in front of us to copy. Dreams are a simulation. Because they appear so life-like, we forget that dreams are akin to incredible, self-generated special effects, produced in the audio-visual centers of our brain. Hands are the most dramatic example, but they are not the only thing we have trouble recreating in our dreams.

There are other ways dreams fall short in how they reproduce reality; giveaways that you are lucid dreaming. Lucid dreaming experts suggest you can push on a solid object to see if your hand goes through it, or check your reflection in a mirror to see if it looks normal.

Another clue can be found in watches or clocks. They, too, seem to be off in dreams. Digital watches and clocks may have no numbers, or the numbers may be hard to read, or they may morph in strange ways. The hands on analog watches or clocks may also move or change in bizarre ways.

Wake-Initiated Lucid Dreaming

A second technique developed to initiate lucid dreaming is called Wake-Initiated Lucid Dreaming (WILD). The WILD technique is designed to jump from waking life straight into a lucid dream and may be the most difficult to master. Researchers suggest you can use it as you're taking a nap, going to sleep for the night, or going back to sleep after waking up.

The WILD technique calls for you to lie down and relax, remaining still and taking slow, deep breaths until you reach sleep-entry. As we learned in the chapter on dreams and creativity, this is the mind-wandering state just before you fall asleep. When you are in this sleep-entry state, try to keep your mind awake as your body falls asleep. To keep your mental vigilance as you drift off, you can try verbal priming by repeating a phrase like "I will have a lucid dream," or "I will become lucid."

Another approach to WILD that has reportedly been successful is counting to sleep: "One, I'm dreaming. Two, I'm dreaming," and so on. Proponents of this approach say you can also focus on the slow inhalation and exhalation of your breath, the imagery in the hallucination, or bodily sensations as you drift off, moving your attention from one part of your body to another in a systematic fashion.

The Tibetan Buddhist practice of *yoga nidra* has used the WILD method for centuries. Practitioners lie down in Shavasana, the corpse pose, and then move their attention around their bodies, relaxing each part of the body in order. They visualize their breathing as they fall asleep, with the intention of maintaining mental awareness as they become lucid. Once lucid, the meditation continues, with the goal of experiencing the divine.

The WILD technique is essentially the opposite of a typical, spontaneous lucid dream. In a spontaneous lucid dream, you are dreaming and realize you're in a dream. In other words, the dream comes first, then you become lucid. With WILD, you are attempting to keep your lucidity as you drift into the dream state.

One experiment found WILD worked particularly well during napping if the dreamer woke up two hours early and then took a two-hour nap either at their normal waking time, or two hours later than the normal waking time. Both of these nap times worked well for bringing about lucid dreams.

The timing with this and other lucid dream techniques is geared toward a typical sleep cycle, which lasts ninety minutes, and is designed to disrupt your sleep just before REM dreaming. The period of REM is shorter earlier in the night, about ten minutes, and gets longer as the night progresses, with the final REM stage lasting up to an hour. This provides the biggest target of opportunity.

Recall from Chapter 1 that people deprived of REM sleep will immediately jump into REM. So it makes sense that the WILD technique works by waiting until near the end of a full night's sleep and waking you up just before the final, longest REM stage of the night. By deliberately skipping the biggest block of REM of the entire night's sleep, the mind is eager to jump straight to REM during the nap period. The phenomenon is called REM rebound. Since lucid dreams typically happen during REM, this strategy boosts the chances for success with WILD.

Using the Power of Suggestion to Lucid Dream

A third lucid dreaming technique researchers developed is called Mnemonic Induction of Lucid Dreams (MILD). This technique combines interrupted sleep with the stated intention of entering a lucid dream. With MILD, you wake up after five hours of sleep and before going back to sleep repeat the phrase, "The next time I'm dreaming, I will know that I'm dreaming," or some other phrase that makes your intentions clear. You can also visualize yourself in a lucid dream.

The strongest predictor of whether the MILD technique will produce lucid dreaming is how quickly you fall back asleep after completing the mnemonic technique. In one study, almost half of the participants experienced lucid dreams if they were asleep again within five minutes. It's not clear why this matters, but it seems likely these dreamers go right back to REM sleep.

If it seems unlikely that simply stating your intentions will affect your dreams, remember that you are the dreamer. Why wouldn't you be able to influence your own dreams? This is similar to priming your dreams to focus on a certain problem, person, or topic by stating your intention out loud or writing it down before bed.

Before the previously mentioned British artist Dave Green goes to bed he uses elaborate rituals to prime his mind to paint in his lucid dreaming. He might meditate for twenty or thirty minutes before bed or pace the room and rehearse the actions he's intending to do in the lucid dream. In a video describing his technique, Green says he places a pen and paper by his bed and writes down his goal for the lucid dream. He says these rituals help him focus on what he's planning to do in his dreams.

A related technique that is often combined with this mnemonic induction technique is called Wake Back to Bed (WBTB). You fall asleep and wake up after five hours, stay awake for thirty to 120 minutes, then go back to sleep. The interrupted sleep makes it more likely you will re-enter sleep in the REM part of the sleep cycle.

Senses Initiated Lucid Dream: 'A Very Mysterious Technique'

SSILD may be the first lucid dreaming technique that was crowdsourced. It was introduced on a Chinese lucid dreaming forum by a blogger who uses the name Cosmic Iron, though in the scientific literature he is identified as Gary Zhang.[2] The name he initially gave the method was 太玄功, which translates literally to "a very mysterious technique." He later called it Senses Initiated Lucid Dream to match the naming convention of other lucid dream induction methods. The second "S" in the acronym is intentional. Zhang's goal was to make a technique that was, in his words, "idiot proof" and required no visualization or creativity of any kind.

Here's how the method works. First, set your alarm to wake you up after four or five hours. When the alarm goes off, get out of bed for five or ten minutes. During that time, go to the bathroom, walk around, but do not do anything too rousing. After returning to bed, lie down in a comfortable position, and cycle through the senses. Focus on sight. Turn your attention to the darkness behind your closed eyelids. Then focus on hearing, though there most likely won't be much to hear. Practitioners who use this technique successfully say they don't actively try to hear anything but listen passively, almost like meditation. Finally, focus on your senses. What do you feel lying in bed? Your body against the mattress. A sheet or blanket. You should be a passive observer to what you are feeling. A key to this technique appears to be not trying too hard.

Perform the cycle quickly three or four times as a warm-up, then perform the cycle slowly three or four times. Take your time, spending at least thirty seconds with each step. As your mind wanders, do not suppress these thoughts. If you get distracted, just return to the beginning of the cycle. When you've finished, return to your most comfortable sleeping position and fall asleep as quickly as possible.

In studies evaluating how well this method works compared to other, more established techniques, SSILD more than holds its own. In fact, researchers have found SSILD works just as well as the other techniques to induce lucidity devised in sleep research labs. In one study, during the first week of trying SSILD, one in six dreamers were able to experience a lucid dream, which was considered a promising result. Interestingly, false awakenings appear to be common with this technique, meaning the dreamer thinks they're awake but are still dreaming.

But how does SSILD work? How could focusing attention on the visual, auditory, and physical senses result in lucid dreams? Like a lot about lucid dreaming, the answer isn't clear. Perhaps cycling through the senses heightens activity in the Executive Network as you are falling asleep. As we learned, during lucid dreams the Executive Network is more active than during normal dreams, when it is usually dormant. Boosting the Executive Network could allow for the self-awareness needed for lucid dreaming.

Another possible explanation is that paying attention to sights, sounds, and physical sensations could function as a sort of reality testing, alerting dreamers when they've entered the dream.

Combined Induction Technique

In an article published in the journal *Consciousness and Cognition*, a German research team headed by Kristoffer Appel were able to get novices to lucid dream in two nights in a sleep lab.[3] The lucid dreams were not self-reported but verified by the L-R-L-R signal. This is a phenomenal success rate for lucid dreams verified by the lucid dreamer within the dream.

Here's how it worked. When the participants had slept for

between five-and-a-half and six hours, and were fifteen minutes into a period of REM sleep, researchers woke them up. This was designed both to increase the chance they would remember the dream they were just having and the odds they would resume REM sleep when they returned to sleep.

Participants stayed up for one hour. During that time, they stayed in bed and wrote down a dream report of the dream they had just been having. Then, they were asked to get up, sit on a couch, and write down the so-called "dream signs" within the dream report. These are aspects of their dream that would be implausible or impossible in waking life.

The participants then categorized the dream signs. Was it because the action was unlikely or impossible? The form? The context? This task took thirty to forty-five minutes. The idea was to get them attuned to elements of their dream that would indicate it was a dream and trigger the insight that produces a lucid dream. The ultimate goal, of course, was that this attentiveness to dreams versus reality would carry over to the next time they went back to sleep.

Before they went back to bed, the test subjects were asked to reminisce about their previous dream. As they did, whenever they came upon a dream sign, they were told to imagine realizing they were in a dream. Finally, they were asked to mentally rehearse, repeating the phrase, "The next time I am dreaming, I will remember to recognize I am dreaming." The participants went back to bed, and the lights were turned off exactly sixty minutes after they'd awoken. They continued practicing this phrase until they fell asleep.

On the first night of the study, five of the twenty participants experienced a lucid dream they verified with the L-R-L-R signal. The next night, five of the remaining fifteen participants had a lucid dream. Remember, these were novices.

Though the technique in this experience is elaborate, it is something that can, for the most part, be replicated at home.

Does the process to induce lucid dreams have to be so elaborate? Saint-Denys didn't need to go through this complex step-by-step process to lucid dream most nights. But if you think about his methods, he was accomplishing many of the same elements as the participants in the lab. He was writing down his dreams. He was considering which parts of his dream were realistic and which could only have occurred in a dream. Doing so, his brain became attuned to be on alert for dream signs that would spark the awareness he was in a dream.

Yet how the attuned mind informs the dreaming brain when it recognizes a dream sign remains unknown. What Saint-Denys wrote almost two centuries ago still rings true, "We know too little about the mysterious ties that bind the mind to the physical."

About a third of people who have lucid dreams are able to control them, and those who are truly experts learn to routinely control the actions in a lucid dream. Flying, talking to characters within the dream, and having sex are the three most popular actions among this top echelon of lucid dreamers. Other popular planned actions include meeting specific characters, sports, and changing the scene or the landscape. As a lucid dreamer controlling the action, you are the producer, director, and star of your drama.

Drugs Help Induce Lucid Dreaming

Aside from different techniques for inducing lucid dreaming, are there drugs or other substances you can take that make lucid dreaming more likely to occur? Psychedelic drugs such as mushrooms, ayahuasca, and LSD are commonly thought to produce a dreamy, surreal experience, but these are not actual dreams. I can say that with confidence because the brain network activation is different in those experiences. Compared to dreaming, the Imagination Network is less activated in psychedelic experiences. That doesn't mean they aren't creative or profound, but they have more overlap with dissociative states, where someone has the sensation of floating outside the body. Psychedelics can generate something called ego dissolution. At their most powerful, they can help cancer patients cope with their diagnoses and offer other mental health applications, but the experience should not be confused with dreaming.

However, there is one drug scientifically shown to increase lucid dreaming: galantamine. Galantamine boosts the levels of acetylcholine in the brain. Acetylcholine is a neurotransmitter essential for memory and thought. In people with dementia, galantamine may improve the ability to think and may slow the loss of cognitive function.

Galantamine also affects dreaming. It reduces the time between the onset of sleep and the first period of REM, which is called REM sleep latency. It also increases REM density, or the intensity of eye movements during REM. Greater REM density corresponds with more intense dreams. Accordingly, galantamine is associated with more bizarre dreams.

To test whether the drug helps induce lucid dreaming, Stephen LaBerge at the Lucidity Institute in Hawai'i did a double-blind study comparing three different doses of galantamine against a placebo.[4] Neither the researchers nor the test subjects knew who was taking the galantamine and who was taking an inert pill. On three consecutive nights, test subjects were awoken after four and a half hours, took the pills and then stayed out of bed for another thirty minutes. They went back to bed and used the Mnemonic Induction of Lucid Dreams (MILD) technique as they returned to sleep.

The results were dramatic. The 4 mg dose of galantamine did twice as well as the placebo, while the 8 mg dose did three times as well. Almost half of test subjects who took the highest dose were able to dream lucidly. When a higher dose results in more dramatic results, it's called a dose-dependent response, which is strong evidence of causality. And regardless of whether the participants had lucid dreams or normal dreams, the drug also increased dream recall, the vividness of dreams, their complexity, and the positive emotions associated with them.

That said, the effect of galantamine was more pronounced when the dreams were lucid. Although we don't exactly know how galantamine increases lucid dreaming, it's possible additional acetylcholine in the brain may enhance activation in the part of the Executive Network that is revived during lucid dreaming.

Indigenous cultures have used supplements and minerals to enhance dreams for generations. In Mexico and Central America, the herb *Calea zacatechichi* is prized as a traditional treatment for a wide range of ailments, from an upset stomach to diabetes to skin diseases, and is also used in dream rituals. In Oaxaca, Mexico, dried *Calea zacatechichi* leaves are smoked as a dream "voyaging" aid by Chontal shamans seeking divine messages—and willing to put up with the potential side effects of loss of balance, retching, and vomiting. In Africa, the Xhosa diviners seek out medicinal roots called *ubulawu* to bring about vivid or lucid dreams. One of them, *Silene capensis*, a fragrant white flower that opens in the evening during spring and fall, is used to induce powerful dreams in hopes of receiving messages from their ancestors.

Lucid Dream Tech

Gadgets like special headbands, eye masks, and smartwatches are now being marketed as ways to assist with lucid dreaming, with more than half a dozen sold commercially. These at-home gadgets are designed to work by pinpointing when the sleeper is in REM sleep. Some do this directly, looking for eye movements, while others use heart rate and accelerometer data to infer REM sleep. Since we are paralyzed during REM, the accelerometer will show no movement. And because we perceive our dreaming activities as real, our heart rate increases. Combine these two data points, and it's possible to determine if someone is in REM sleep.

Once these devices know you're in REM sleep, they try to produce subtle cues, dream signs, that will alert you that you are in a dream. These devices use haptic signals like a vibration, an audible cue, or a visual one like flashing lights. One of these devices will even play a recording of your voice saying, "I am dreaming." If the signaling works, these vibrations, sounds, or lights could make it past the brain's thalamic gating (which shuts out most external cues during sleep) usually without waking you up. The signals serve as cues that trigger lucidity—and can be incorporated seamlessly into the dream as you enter the world of lucid dreaming.

Before these gadgets were widely available, similar signaling devices were tried out in sleep labs. In a test of how well lights would work to induce lucid dreaming, they were used on test subjects on alternate nights without their knowledge, to eliminate any potential placebo effect. Of the lucid dreams reported, two-thirds happened on nights with the light cues.[5]

For the signals to work, however, it helps if dreamers mentally prepare for them beforehand. In sleep labs, test subjects are shown the cues before they go to sleep. It could be a faint flashing light, or a few notes from a violin. When they receive this signal, they are asked to perform a reality check: Am I awake or am I dreaming? Then they are told to be critically aware and notice if their experience is different from a normal waking experience.

Normally, arousal during sleep originates in the brainstem. When we're sleeping, the thalamus alerts the Executive Network if we need to wake up. It is this bottom-up internal screening pathway that is skirted with devices that flash signals, alerting you that you're in REM sleep. The signal could slip past the thalamic gating without waking us up.

But what if you could reverse engineer the body's arousal mechanism? What if instead of going bottom-up, you went top-down?

Researchers are trying to do just that using non-invasive brain stimulation techniques. Transcranial stimulation, which we learned about earlier (see page 133), has shown it may increase dream self-awareness, though there is scant evidence it can produce lucid dreams—at least, not yet. As our understanding of the neurophysiology of lucid dreaming expands, it seems reasonable that researchers will eventually dial in the correct frequency and precise locations in the brain to stimulate lucid dreaming.

The lack of results thus far has not slowed the hunt for a non-invasive way to reliably spark lucid dreams. Researchers across the world are racing to find it. Sérgio A. Mota-Rolim, at Federal University of Rio Grande do Norte in Brazil, and his colleagues have argued that there may be more than one point of entry into lucid dreaming, with each door leading to a different experience: first-person control, third-person body imagery, or enhanced visual vividness.[6] At the time of writing, the elusive key—or keys—have yet to be found.

By and large, lucid dreaming is viewed as a positive experience that offers unique opportunities for creativity, problem-solving, and even practicing skills that lead to improvements in life. Lucid dreamers say the experience boosts their waking mood and they feel refreshed the following morning. However, it's important to keep in mind that most techniques to induce lucid dreaming involve forced awakenings and interrupted sleep. By definition, inducing lucid dreaming with the Wake Back to Bed or similar techniques fragments sleep and ultimately disrupts the sleep architecture. It also has the potential to decrease the total amount of sleep if the lucid dreamer is not careful. At the same time, lucid dreaming can transport you to a truly unique state of consciousness, the surreal intersection of dreams and self-awareness.

8.

The Future of Dreaming

For two decades, Japanese researcher Yukiyasu Kamitani has gotten closer and closer to being able to decode a dream and turn it into a video.[1] Beginning with a computer algorithm that could decode data from brain scans to determine if someone saw a pattern of lines vertically, horizontally, leaning left or leaning right, Kamitani and his team can now tell you with confidence what you were dreaming about just before you woke up. Was it a person? A tree? An animal? His computer algorithm is sophisticated enough to know.

This was no easy feat. To recreate visual images based only on real-time blood flow in the brain and electrical activity on the brain's surface, Kamitani and his colleagues at Kyoto University capture brain activity represented by voxels, or three-dimensional pixels, and process them using a deep neural network, a type of machine learning able to execute incredibly complex computational tasks. With a deep neural network, processing all this information becomes more efficient over time as the computer finds patterns in the vast amount of data. Using a high-powered computer, the information is then pieced together by a reconstruction algorithm.

Kamitani has gathered a lot of dream data by putting people into an fMRI to capture real-time metabolic activity in the brain; simultaneously, an EEG records electrical activity. The participant is repeatedly woken up just as they are falling asleep, the visually rich sleep-entry state when the mind starts wandering freely. Each time a test subject is awoken, a lab technician asks if they saw something just before they woke up. The test subject might say they saw a plane, a girl, or a black box. These images are matched up to the brain activity occurring at the time, then the test subject is asked to go back to sleep. Do this enough times, and the machine learning algorithms start finding correlations between what's happening in the brain and the images the test subjects report.

Taking advantage of enormous advances in artificial intelligence, other researchers around the world have joined this effort to translate brain activity into visual images. As a result, decoding neural signals is becoming increasingly accurate. It is entirely conceivable that in the next decade or so, we will be able to take the brain activity of someone dreaming and translate it into a visual reproduction of a dream.

At Jack Gallant's cognitive neuroscience lab at the University of California, Berkeley, for instance, he and other researchers have in the last decade been able to decode the brain activity of people watching movie trailers.[2] From brain imaging alone, they are able to decipher with an astonishing degree of accuracy what the viewer is watching. The brain activity of a test subject watching a movie trailer for *Bride Wars* was correctly labeled as a woman talking.

Instead of relying on a three-dimensional map of the brain for analysis, however, Gallant flattens it out, the two hemispheres looking something like mirrored images of a map of Australia. He tracks 100,000 points on the cortex on his map, looking for relationships between what the brain is doing and what the person is watching. Specifically, he focuses on the visual cortex, which is represented near the very center of his flattened brain map. More than average brain activity is red. Less is blue.

Gallant's lab has started decoding the mind by listening to stories or reading transcripts of those stories. Using fMRI data, they have been able to create a functional map, correlating concepts in the stories with specific brain activity. This isn't as simple as putting a pin on a map, however. Every concept activates dozens of areas of the brain. Despite the challenge, researchers in his lab can now discern, based on brain activity alone, whether someone is reading or listening to something involving time, a place, a person, a body part, or a family relationship, whether it is tactile or violent, whether the story is focused on visual information like texture or color.

What's fascinating about this elaborate mapping is these are the same sort of semantic connections the dreaming mind follows.[3] When you think of an object such as a car, you might think of the current car you own, what you know about the history of cars, or how to drive a car. Maybe you'll think of the car you learned to drive in. Or you might think of other modes of transport, or taking car rides with your father or mother as a child. Depending on what you think of, a different area on the brain will light up related to procedural memory, episodic memory, semantic memory, and affective memory.

Nevertheless, we still have a long way to go before dreams can be decoded with precision. One challenge is that everyone's brain is a little bit different. This is something I see in the operating theater all the time. The fine structures of the brain may be in roughly the same area, but there are always slight variations. When it comes to decoding brain activity or engineering brain activity, there would need to be some sort of standard way of calibrating an individual's brain against a general map.

Another challenge relates to the technology itself. Because fMRI machines capture images more slowly than, for example, the twenty-four frames a second of films, the decoded images lack continuity. This will no doubt change with time, but for now, MRI machines most often sample only 2.5 times a second.

They also lack the necessary resolution. The typical MRI in clinic is 1 tesla, a measure of magnetic strength. The one used by Berkeley researchers is 3 tesla. But even a 3 tesla MRI can only measure brain tissue down to a 2 millimeter cube, which is the basis for the data Gallant's lab uses. Unfortunately, the area is imprecise when it comes to looking at brain function. It's like having a satellite view of a neighborhood rather than a particular street. The next generation MRI scanner should be able to scan down to a .4 millimeter cube, or 400 microns, which will allow for much more precise brain mapping.

If dreams can someday be decoded from brain activity, the question then becomes: Will we someday be able to do the opposite? Will we be able to engineer a dream from nothing? Will we be able to choose a dream the way we choose a movie from a streaming service? The idea may sound like science fiction but someday, and perhaps sooner than we think, it may be possible.

Dream Engineering

In the first half of the twentieth century, most people said their dreams were in black and white. This was also a period when newspapers, photographs, television, and most movies were black and white. Color dreams were thought to be the exception and were dubbed "Technicolor dreams," named for the process that produced color movies beginning in the 1930s.

In the 1960s, all that changed dramatically. Most people

began reporting that they were dreaming in color. The catalyst? A decade earlier, there was a massive shift in the media from black and white to color. The first commercial color TVs went on sale. Magazines, too, made the switch from black and white, and movies were beginning to be shot in color. This change in dream reports appears to be a byproduct of how popular culture transformed in the last century.

What if you set out to change the look of dreams? Could the dreamscape be engineered? Researchers have tried to manipulate the appearance of dreams, with limited success, as we've seen in experiments in which test subjects wore colored goggles or played immersive video games. The dreamscape changed, but not completely and not in any predictable way. The dreaming mind appears to be too wild to be corralled in this way.

If engineering a dream's "video" is difficult, how about the "audio?" Can we manipulate what someone will hear in a dream? It appears the language we hear during the day can have an impact on our dreams. In studies of people who are bilingual, the language of pre-sleep interviews influenced what language the test subjects dreamed in. Similarly, researchers found that some English-speaking Canadians taking an intensive six-week French class started dreaming in French. Like the visual aspects of our dreams, these studies show that what we hear during the day influences our dreams. But trying to predictably manipulate our dreams with audio cues, while asleep, is still being explored.

Interestingly, it is not sights or sounds but smells that may offer the most potential in the short term for some measure of dream engineering.

Using the Senses to Influence Dream Content

As we've learned, when we're dreaming, the outside world is shut off—but not completely. One way into our thoughts and dreams is through the least regulated of our senses, the sense of smell. The sense of smell is connected directly to the parts of the brain linked to the memory and emotional systems of the brain, the hippocampus and the amygdala.

Smell has another feature that makes it ideal for dream engineering: It bypasses the thalamic gate that blocks most sensory signals from reaching you as you dream. This may have had evolutionary advantages. Smelling fire or the scent of an animal nearby while you were sleeping could have been a lifesaver in prehistoric times.

Thanks to the lax thalamic gating for olfactory stimuli, some smells can still affect dreams without waking up the dreamer—and without the dreamer knowing it. The smell of rotten eggs can turn dreams negative. The smell of roses makes pleasant dreams more likely. Of course, there are limits. If the scent is too powerful, it will pierce the veil of sleep and wake up the dreamer.

Scents can also be used during sleep to help you learn. If you smell the scent of pine while you study a new language, having a device that releases the same scent while you are asleep appears to promote learning by strengthening the memory. In one study by Laura Shanahan at Northwestern University, test subjects tried to remember the placement of different categories of pictures on a grid.[4] The pictures showed animals, buildings, faces, and tools; each was assigned a corresponding scent. For example, the smell of cedar might go with any picture of an animal, while the smell of roses went with pictures of buildings. During sleep, the participants would be given only some of the odors. When they went to take the memory test after they woke up, test subjects had a better memory for pictures reactivated by odors as they slept, though they had no idea why.

In some studies, researchers have found that targeting the delivery of odors during sleep and dreaming may even have the power to fight addiction. In one study, sleepers exposed to the combined smell of cigarettes and rotten eggs smoked 30 percent fewer cigarettes the following week.[5] This type of olfactory manipulation works both ways. A whiff of smoke during sleep causes cigarette smokers to smoke more the next day. Interestingly, the ability of smell to influence behavior seems uniquely tied to sleep. Combining cigarette smoke and rotten eggs for test subjects who were awake did not produce any results.

Given that smartwatches can now detect which sleep stage we are in, I can imagine them synching with devices that emit odors to help with learning or with therapeutic goals. The technology seems straightforward enough. Smells could even be used to manipulate the content of our dreams, something pioneered more than a century ago in France.

Léon d'Hervey de Saint-Denys, whom we met in the previous chapter, wanted to see if he could target specific memories in his dreams through scents. To test this hypothesis, the nineteenth-century Parisian purchased a different perfume each time he traveled. He'd douse a handkerchief in the perfume, and each day in a particular place he'd smell the handkerchief. Returning home, he'd wait a few months and then have a servant shake a few drops of the perfume onto his pillow. As a result, he'd dream of the place he'd been when he'd smelled that perfume. Taking this a step farther, he started having the servant put drops of two different perfumes on the pillow. Amazingly, he described combining elements of both trips in his dream.

With this informal experiment, Saint-Denys reported that he could engineer his dreams. The scents he had connected with certain memories during the day were reactivated by the same olfactory cue as he slept. His design to steer his dream in a certain direction was more scientific but little different from the types of dream incubation that have happened for millennia.

It's not only smell that has been used to boost the learning process. Musical themes can be used in the same way. In one study, participants attempting to solve a puzzle heard the same musical theme repeated over and over. Those who had the music played quietly as they slept were more likely to find the solution to the puzzle in their dreams than those who didn't.

Tactile cues can also change dream content. Touching a dreamer's leg to generate a reflexive knee jerk when they are dreaming can prompt dreams of falling. Put a dreamer's hand in water, and the dreamer is more likely to incorporate water into the dream narrative. In fact, spray a dreamer with water, and water makes its way into almost half of dreams: A dreamer might dream of being in the rain or swimming.

There are other ways to direct the content of your dreams, though I wouldn't necessarily recommend them. If you deprive yourself of water, you're much more likely to dream about thirst or water. If you watch a stressful movie right before going to sleep, you are more likely to have a dream that is negative than one that is positive. Of course, the opposite is likely true as well. As we've learned, one way to lessen the chance of nightmares is a calming bedtime ritual.

The Insidious Future of Dream Advertising

The advertisements we see during our waking hours are an overt attempt to influence our thinking. Advertisers are now taking aim at our dreams. What makes dream advertising potentially far more malignant is that it happens outside our conscious awareness. As we've learned, when we're dreaming our rational brain is offline, which means we are less skeptical and more vulnerable to targeted messaging. One study has already found dreaming about an ad makes us more likely to buy that product.[6]

Even with the current limits of dream engineering, companies are already getting into the business of targeted dream incubation. In their view, apparently, dreams are the last great, untapped landscape for marketing their products.

In 2021, Molson Coors Beverage Company attempted to use targeted dream incubation to infiltrate the dreams of consumers leading up to the Super Bowl, the American football championship game. The company couldn't advertise its beer during the big game because the league had an exclusive contract with a competitor. A marketing vice president came up with an end run: If the company couldn't run an advertisement during the game, could it run one in people's dreams?

Molson Coors approached Deirdre Barrett, the Harvard dream psychologist. Company executives wanted to know if they could create an ad that had the power to infiltrate dreams. The goal was to firmly plant the ad in people's subconscious and have it play in their dreams. Barrett told them it was possible to influence dream content, but only if there were cooperative subjects.

With Barrett's guidance, Molson Coors produced a trippy,

visually intense, ninety-second advertisement and called it "The Big Game Commercial of Your Dreams." They also released an accompanying eight-hour soundtrack. In the film, as dreamy music plays, a translucent avatar flies through mountains and over a stream, interspersed with images of the company's products, and other images of nature, cartoon figures, and mesmerizing shapes and patterns. The color-saturated video moves quickly from place to place and shifts from surreal images to abstract shapes and objects, much like a dream.

Testing it in a sleep lab, subjects were shown the ad several times and then told to prime their dreams by saying, as they fell asleep, that they wanted to dream about the video. When they were woken during REM sleep, the subjects reported dreaming about a waterfall, or walking through snow—images from the video. One dreamer, her voice groggy because she'd just woken up, said the mountain in her dream had something to do with Coors beer. In fact, five of the eighteen test subjects reportedly had dreams that incorporated some element of the advertisement.

Molson Coors posted the video online, inviting consumers to watch it and participate in what the company called "potentially the largest sleep experiment ever." They suggested that people watch the video several times before going to sleep and play the soundtrack as they slept. Participants were offered discounts and urged to post their dream reports on social media using a hashtag and tagging Coors Light and Coors Light Seltzer. The company claimed the advertisement was an enormous success, with 1.4 billion impressions, a 3,000 percent increase in social engagement, and, maybe most importantly for the company, an 8 percent increase in sales.

Dreams, the once sacred and sacrosanct province of the dreamer, are now a target for marketers, and Molson Coors is not the only interested player. In a 2021 "Future of Marketing Survey" by the American Marketing Association, 77 percent of 400 companies said they planned to experiment with dream advertising by 2025. It appears a gold rush is underway to manipulate the fertile terrain of our dreamscape.

Burger King sought a different path to hijack the dreamscape. As a Halloween promotion, Burger King introduced the Nightmare King burger, the tagline being: "Feed your nightmares." The sandwich included a burger, a chicken patty, bacon, cheese, and a bright green bun. Aside from the hefty number of calories, the only truly unusual thing about this sandwich was the brightly colored green bun, yet the company claimed the Nightmare King was somehow effective in inducing nightmares.

To prove the burger's nightmare-inducing properties, Burger King partnered with a diagnostic and sleep lab, which tracked the dreams of 100 participants over ten nights. According to a news release from Burger King, the Nightmare King more than tripled the incidence of nightmares. Of course, the mere suggestion that eating a certain meal could cause a nightmare is potentially enough to cause more nightmares.

What's interesting is that Burger King's Nightmare King was a cheeseburger, and cheese has long been (falsely) believed to cause nightmares. In Charles Dickens' *A Christmas Carol*, Ebenezer Scrooge at first blames "a crumb of cheese" for the appearance of the ghost of his former partner, Jacob Marley. There's no proof cheese can or does cause nightmares, but the belief is enough to keep the myth alive. This self-fulfilling negative consequence is akin to the nocebo effect, which is the opposite of the placebo effect. If you believe a medication will cause certain side effects, it is more likely to cause them.

These nascent efforts to get you to dream of beer or burgers is likely just the beginning. The night is fast approaching when advertisers routinely target your sleep and dreams, trying to influence your waking behavior while your guard is down, threatening to infect something so vital to our wellbeing. The sacred refuge of sleep and dreams may soon be under siege.

This possibility has the research community more than a little concerned. In an open letter written in response to the Molson Coors ad, thirty-eight researchers from around the world argued against letting dreams become another playground for corporate advertisers and expressed their support of legislation prohibiting advertisers from targeting people as they slept. Responding to the Molson Coors campaign, they asked: "What have we lost when we become so collectively inured to invasions of our privacy and to exploitative economic practice that we would accept a twelve-pack for the placement of beer advertising in our dreams?"

Technology and Dreaming

We've learned that flashing lights, vibrations, heating and cooling the air around the dreamer's skin, and audible cues can all be used to target specific memories. An early test of this even found verbal cues of words related to liquid during dreaming sleep caused an increase in related dreams—and they can affect the subject's behavior when they awaken.

For example, verbal cues have also been used to influence nappers' brand preferences. In a study by Chinese researchers Sizhi Ai and Yunlu Yin, participants heard one of two brand names over and over as they slept. When they awoke, they were more likely to pick the brand they'd heard while they were sleeping. Sizhi Ai concluded that the "neurocognitive processing during sleep contributes to the fine-tuning of subjective preferences in a flexible, selective manner."[7] A control group was given the same repetitive message without napping, but it had no effect. How this works is not known, but changes in the sleeping participants' brainwaves signaled when they had been influenced.

Based on this and similar studies, could a smart speaker, smartwatch, or some other external device or app offer you shopping cues while you sleep? It certainly seems that way. Already, smart speakers have infiltrated our bedrooms, and smartwatches and other devices are now able to monitor our sleep cycles. Based on movement, heart rate, and other signals, these devices have a pretty good idea what stage of sleep you're in; the Apple Watch will even track your REM sleep.

Given that we are vulnerable to audible cues while we're sleeping, will the end user license agreement (EULA) for our smart speaker or wearable tech in the future include the right for a company to send you quiet audio marketing messages while you sleep? Will we need to pay extra for ad-free dreaming? And if companies can use devices to infiltrate your dreams, is there anything stopping governments from engaging the sleeping minds of their subjects with propaganda and other mind control? This sort of dark speculation evokes such science fiction as George Orwell's *1984* or Philip K. Dick's sci-fi novel *Do Androids Dream of Electric Sheep?*

The future might involve a more direct machine–brain interface. Right now, people with epilepsy can have a device implanted in the brain that will monitor brainwaves looking for the unique signature that precedes a seizure, which it then disrupts with counter currents. This is a closed-loop circuit, mind and machine working seamlessly and autonomously. Will we be able to opt for surgery to implant a device that will modulate dreams on demand? It sounds radical, but what if it broke the spell of recurring nightmares? Would that be worth an elective surgical procedure? How about if the device gave you more creative dreams? What if you could induce erotic dreams whenever you wanted one?

In the movie *Inception*, ideas were smuggled into people's dreams. In reality, neuroscientists could use implants right now to trigger specific memories. These could be personal memories, but also memories of a specific product. There is also a generation of non-invasive, brain-user interface devices on the market. Is anything stopping these companies from adding a marketing component to their consumer products or misusing the neural data they collect?

The issue has captured the attention of the United Nations Educational, Scientific and Cultural Organization (UNESCO). In July 2023, the organization assembled neuroscientists, ethicists, and government officials to discuss possible regulations to address neural rights. A UNESCO report published at the same time said neurotechnologies will potentially develop the power to access our minds, alter individual personalities and behaviors, and change recollections of past events: "This calls into question fundamental rights like privacy, freedom of thought, free will, and human dignity."[8]

Other organizations have started working to protect people from potential misuse of neurotechnologies. The Neurorights Foundation, founded in 2017, is pushing governments to pass laws to keep private any data gathered by neurotechnologies such as smartwatches, earbuds, and headsets, to limit the commercial use of this data, and to protect individuals from external manipulation. This would include attempts to manipulate dreams. Rafael Yuste, a Columbia University neuroscientist who co-founded the organization, said companies in the fast-growing field have adopted a predatory attitude toward brain data. In fact, the Neurorights Foundation identified eighteen neurotechnology companies that required users of consumer devices to relinquish ownership of their own neural data.

Governments are starting to pay attention, too. In 2021, Chile became the first country to change its constitution to protect brain activity and information. Other countries are considering legislation, but unless it is truly a global effort, it will be hard to protect individuals from the potential abuse by neurotechnologies. As Yuste put it in an interview, "This is not science fiction. Let's act before it's too late."[9]

As individuals, we can take steps to protect the sanctity of our dreams. We can sleep in an environment free of potential messaging, whether it's from our phones, smart speakers, or other devices. We should avoid neurotechnologies that have user agreements giving companies control of our neural information. Dreams can provide us with rich insights, revealing much about our emotional state—I think it's important to keep them unsullied by commercial interests.

9.

The Interpretation of Dreams

Researching and writing this book has led me to see not only dreaming, but neuroscience itself in a new light. In the practice of medicine and surgery, I've witnessed the power of dreams to persist in the face of terrible injury. I've seen children who have had half their brains removed as a last-resort treatment for intractable seizures still report dreaming. Dreams make themselves heard.

More than that, dreams are especially relevant because they provide a form of thinking and feeling only possible through a unique set of neurochemical and physiological changes. It is only through dreaming that we have privileged access to this mental space. We couldn't think this way during our waking hours if we tried.

This is why dreams are worth paying attention to. They can give us insights we wouldn't achieve otherwise. They can make associations between people from different times in our life, between seemingly unrelated events, between what's happened in the past and what may happen in the future. It is because of the powerful underlying neurobiology of dreaming that I am convinced that dreams have meaning and purpose. And that makes reflecting on dreams an important aspect of a life lived fully, a life examined. At least, I know it has become so for me.

You may have thought someone who has spent a career

immersed in the brain would reject dream interpretation as nothing more than pop psychology, akin to reading a horoscope. When I began researching and writing this book, I might have agreed. However, given the rigorous science behind our understanding of what happens in the brain when we dream, I now believe dreams can be interpreted. But how?

Websites abound with dream dictionaries saying that if you dream of X, it means Y. Books, too, offer one-size-fits-all answers to the meanings of certain dreams. This type of approach is little different from a dream book written in ancient Egypt more than 3,000 years ago, which listed 108 dreams and their interpretations. Dreaming of the moon was a good thing, meaning the gods were forgiving you. So was a dream of eating crocodile meat, which meant you would become a public official. But if you saw yourself in a mirror in a dream, it was a bad omen, a signal you would soon have to find another spouse.

Dream interpretation was regarded by ancient civilizations in Mesopotamia, Greece, and Rome as an art requiring intelligence and, sometimes, divine inspiration. Not surprisingly, they attributed great significance to dreams, believing them to be communications from the gods or the dead. Dreams were imbued with the power of prophecy, and those able to interpret dreams were held in great esteem. This belief in the power of prophecy is alive and well. Surveys have shown two of every three people believe in the power of dreams to foretell the future.

Freud was a modern descendant of these dream interpreters. In his view, dreams were not messages from a god or the afterlife, but from the subconscious, revealing our repressed desires. The heyday of Freudian psychoanalysis has come and gone, but belief in the power of dreams to give us important information is alive and backed by the sophisticated tools of modern neuroscience.

I am not an outlier in my belief that dreams are a valuable source of self-knowledge. Neuroscientists and psychologists increasingly believe there is much we can learn from our dreams. Studies have shown that interpreting our dreams really can inform our waking lives, though not always in the ways we are expecting.

Why Dream Dictionaries Can't Work

Go online, and it's easy to find the purported meaning of not just your dream but any dream. The internet is full of websites offering to interpret them. What does it mean to dream of a leaf? One website says it's a symbol of change. Just as the leaves change with the seasons, something is ending, and we are starting again. Another says it's a sign of renewal. A third says it's a sign of growth and openness. They all make sense on some level, so which interpretation is correct?

Dream websites cleverly offer a mix of vagueness and specificity that make it easy to shape your personal circumstances to fit any of these interpretations. Isn't something always ending or beginning in your life? Don't we all want to be associated with renewal, or growth and openness? It's human nature to take generic descriptions like these and personalize them. Horoscopes do the same thing. We see a fuzzy description and make it fit our particular situation.

The truth is, the same dream image can mean so many things, not just from one person to the next, but to former versions of ourselves, at different stages in our lives. I had a dream recently where I was walking over a bridge. Look up the meaning of bridge online and you find the same interpretation as dreaming of a leaf. On one website, a bridge symbolizes, "the transition from one state to another, like a rebirth." Another says it's a spiritual message that it's time to review your life, or a sign that most difficulties can be overcome. A third suggests a bridge means there will be a transition in your life. As a metaphor, a bridge can suggest many things: a marriage, two sides coming together, a path to end suffering for those with untreatable cancer.

Just as your waking mind is the unique product of your memories, your day-to-day experiences, and your emotional state, your dreaming mind is, too. Although there are some dreams experienced by many, like falling, arriving late, or being chased, dreams are personal. Your dreams are the product of your brain at this particular moment in your life, and they change with the seasons of your life. To expect them to fall in line with others because they share the same central narrative, or the same visual element, is simply not realistic.

But there is also a neurological reason for why the same image in a dream could have varied meanings to us individually. As we've learned, the mPFC, in the frontal lobes, adds meaning to our experience. Each of our mPFCs serve a similar function, but only with the material inside our minds. When we dream, we're taking different sights, sounds, memories, and emotions and synthesizing them into something that is personally meaningful. The brain is providing the content, and the mind is providing the meaning.

That meaning was created by you and is specific to you. For that reason, you can interpret dreams, those dreams that are your mind's voice, but there is only one person who can serve as the interpreter—you.

The Five Dream Narratives

Dream narratives can take almost infinite paths, shaded by the full spectrum of human emotions. I believe dreams, in general, fall into five manageable categories. When I am trying to interpret a dream, I begin by deciding which of the five dream types I am reflecting on. Each deserves a different approach. Let's go over them one at a time.

Overt Dreams

First, there is a dream whose meaning is overt. If we're taking a test the following day, and we dream the alarm didn't go off, the meaning is clear. This is easily interpreted: Stress about the exam has triggered the dream. The same could be said for dreaming about giving a speech naked or missing an important flight, when both are impending events in your waking life.

Genre Dreams

Second, there are what researchers have dubbed genre dreams. These are dreams tied to a stage of life that changes us in profound ways. Genre dreams are so transparent in their meaning they too need no real interpretation. Two distinct categories of genre dreams are pregnancy dreams and end-of-life dreams.

As you might expect, dreams of pregnant women are more likely to center on themes related to being pregnant, childbirth, bodily anatomy, and being a mother. Women in the final months of pregnancy are more likely to have specific dreams about the baby and its sex. Are they accurate? The scientific literature does not have a clear answer to the question. While one study found all eight women who dreamed the sex of the baby were correct, another found women did no better than flipping a coin.

Pregnant women also report communicating with their baby in their dreams, even of the baby announcing its name to its mother. These so-called "announcing dreams" have a rich history in traditional cultures. Among the Ese Eja in the Peruvian Amazon, for instance, it's tradition for women to dream the names of their children. In these dreams, animals interact with the dreamer to reveal a child's "true name."

After giving birth, the anxiety, stress, and sleep deprivation of new motherhood often result in negative dreams and nightmares. One common nightmare among new mothers has been dubbed the "baby in bed" nightmare: The infant is lost somewhere in bed, suffocating. This prompts the mother to make a frantic search in the covers for the missing child. Once the mother is fully awake and realizes her child is not trapped under the covers, she will frequently still feel compelled to check on the baby.

Another type of common genre dream occurs for people nearing death; end-of-life dreams. People report vivid dreams of dead family members, pets, or other members of the family. For the dreamers, these dreams are often a source of hope and comfort, joy and serenity. They bring peace and acceptance and can lead to the dreamer getting their affairs in order, and reconciling with family members.

Dream reports gathered at a New York center for hospice and palliative care found common themes in the end-of-life dream narratives. These included dreams with a comforting presence. One woman dreamed of her dead sister sitting beside her bed. A man nearing death dreamed of his long-dead mother soothing him and saying, "I love you." The dream was so true to life he could smell her perfume. Others dreamed they were being watched over in their final days. One reported her husband and dead sister joined her for breakfast; another that her father and two brothers, who were all dead, silently hugged her as they welcomed her to join them.

In their final days, other hospice patients had dreams they were preparing to go somewhere, or of dead relatives and friends waiting for them. Three days before she died, one woman dreamed she was at the top of a staircase. Her dead husband stood at the bottom, waiting for her. Most of these dreams were comforting, even if some patients said they were not ready to die.

People who are grieving often also report having dreams of a dead loved one who usually appears at peace, healthy, and free of pain or illness. These dreams are seen as deeply meaningful, spiritual experiences, and they bring acceptance of the loss, a sense of comfort, and lessen grief.

Universal Dreams

The third kind of dream is the universal dream: the nightmare and the erotic dream. As discussed in Chapter 2, children who have not experienced trauma have nightmares not as the result of some pathology but as part of the mind's maturation. Because nightmares often reflect our mental state, adults with anxiety and depression are more prone to nightmares. New-onset nightmares can serve as a thermometer gauging our wellbeing. They can alert us to our emotional state of mind. As we've seen, nightmares related to trauma may offer a window into how well we are processing what happened to us. Dreams triggered by trauma are often replays of the event itself—or something close. The more metaphorical a dream is following trauma, the better the dreamer is thought to be emotionally processing the traumatic event.

Like nightmares, we all have erotic dreams at some point in our lives. As we learned in Chapter 3, many of our erotic dreams are simply the product of the imagination unleashed, without judgment. Infidelity dreams do not signal unhappiness in a relationship, nor do they necessarily suggest an attraction to the focus of your dream's desires. What is more telling is your reaction when a partner has a dream like this. Infidelity dreams that are upsetting to hear are less about the dream and more about the strength of the relationship.

Unemotional Dreams

The fourth type of dream is an unemotional dream. Unless you can point to a strong emotion attached to a dream, finding meaning in this kind of dream may be difficult. I am talking about emotion felt by the dreamer, not any explicit mention of an emotion in the dream. It turns out, emotions are rarely talked about in dreams.

If you remember a dream but attach a neutral or weak emotion to it, in my opinion the introspection is not worth the effort. You wouldn't spend any time analyzing humdrum moments of your mental life when you're awake, so there's no point doing it for a dream. Go after the ones that stir you.

In this vein, some dreams are a jumble of images or events or characters, and they may be emotionally neutral or unclear. These dreams are the equivalent of mental static, no different to the abundant and haphazard accumulation of thoughts during the day. I think these dreams, too, are not worth interpreting.

Emotional Dreams

That leaves the fifth and final dream type, which I believe provides the richest source of insight. These are emotional dreams that have a coherent narrative thread and often a distinct central image. This is the one that's going to take effort to interpret because unlike the first type of dream—where the narrative is overtly tied to something in your waking life—this dream may have a narrative disconnected from your reality.

By focusing on emotional dreams, you focus on the dreams that matter to you. Remember, dreams can bring us to emotional heights impossible in our waking lives. So it shouldn't come as a surprise that they can ripple through your waking mood. All of us have awoken sad, anxious, or elated after a particularly moving dream. Maybe we've awoken thinking about the dream or find ourselves thinking about it during the quiet moments of the day. Sometimes, dreams are simply impossible to ignore. I believe these are the dreams that demand an attempt at interpretation. Dreams like these can provide a portal to your deepest psychological world.

But before we learn how to decipher these dreams, a caveat: There is no way to objectively prove if a dream has been interpreted correctly. We cannot put you in an fMRI to get brain imaging to see if your interpretation matches up with some objective reality. Nor is there a blood test or some EEG reading that could reveal the answer.

To interpret a dream, you first need to remember it. As we learned to do earlier, before you go to sleep, make an autosuggestion that you will dream, that you will remember your dream, and that you will write it down. When you wake, before thinking about the day ahead, write down what you can remember about your dream. You can also record your dream report on your phone. Just make sure it's the first thing you do. Don't check emails or social media first. Most of us have the experience of trying to recover a dream, only to have it slip away. At first, you may only be able to remember a few snippets. If you make writing down your dreams a daily practice, it will become easier and your dream recall will increase quickly over time.

Because you're recording your dream in the morning, you're most likely recalling the dream from the final REM cycle of the night. As the night progresses, dreams shift from being more continuous with waking life early in the night to longer, more emotional and hyper-associative dreams later in the night. The British researcher Josie Malinowski found the final REM dream cycle before we wake up is the most emotional, the most symbolic, and the one holding the greatest personal importance.[1]

How to Interpret Your Dreams

Interpreting your dreams requires keeping in mind how dreams are made. As we learned, dreams are nightly shifts in brain activations and neurochemicals that result in highly emotional and highly visual narratives, marked by novel thinking. These emotions and the visual connections are our own. The dreams are decipherable by us because we are the ones who conjured them.

To make sense of dreams, I have adopted a two-step approach informed by these central aspects of dreaming, and focusing on the emotional and visual aspects of the dream. I'm choosing these two elements—the visual and emotional—because they can achieve an intensity when we're dreaming that is impossible at other times in our lives. This is an approach pioneered by the late Ernest Hartmann.[2] Recent neuroscience has validated this approach, in my opinion, given the brain activation that occurs during dreaming and the patterns that emerge when thousands of dream reports are analyzed.

To use this method, first, look at the dominant emotion and the emotional intensity of the dream. Was it anger, anxiety, guilt, sadness, helplessness, despair, disgust, awe, hope, relief, joy, or love? How intense was the emotion? Sometimes dreams will produce not one, but many emotions. Focus on the strongest emotion in the dream. The more intense the emotion, the more important the dream.

Underlying emotions and emotional concerns shape and drive the dreaming process in our brains. Given the hyperaroused state of the emotional, limbic system during powerful dreams, I believe the dream's dominant emotion is guiding the broad and often irrational associations we make in our dreams. If you are stressed or anxious, your dreams will likely reflect this emotional state, and you're more likely to have disturbing dreams. The images and plot that accompany these dreams can match the emotion while having little to do with the source of the stress or anxiety. This is why fear of starting a new job might elicit a dream of a hike along a dangerous mountain path, or why stockbrokers during a market crash didn't dream of money or stocks but had a spike in dreams where they were falling or being chased.

The second step is to consider the central image of the dream. Like emotions, the visual centers of the dreaming brain are robustly activated. Dreams link images with emotions as a way of contextualizing them. When you consider the central image of the dream, think of it as a metaphor, an image that serves as a symbol for something else; it's important to remember that dreams are another form of cognition, so while they are often bizarre, they are potentially illuminating in a way that is impossible to achieve by other means. For instance, a sexual assault survivor may dream of being swept up in a tornado, an image that evokes the same sort of fear and helplessness as the attack. In one case study that illustrates this point, a man scheduled for major heart surgery had a dream that a quarter of beef (a quarter of an entire animal) had been delivered, and he, his daughter, and an ex-boss were deciding how to cut it up to preserve it. It's hard to interpret this as anything but a dream about his impending surgery.[3]

Often, it seems our dreaming mind searches for other times when we experienced the same sort of emotion and conjures images from that experience. Vietnam veterans going through the stress of marital troubles years later were more likely to dream about the war. For these veterans, the emotion of the dream was key to understanding the dream; the war was a metaphor for the current state of their marriage.

Other major life events can also produce strong emotions and corresponding contextualized images. Dreams recorded after the September 11 terrorist attacks were not about planes or the World Trade Center, but narratives of being threatened in other ways. The COVID-19 lockdowns were less likely to produce dreams about viruses or pandemics and more likely to have dream narratives where the dreamer is trapped, in one example, in a supermarket that turned into a labyrinth.

The scientific literature recounts the dream reports of two women a week after their mother's death.[4] One dreamed of an empty house, without furniture, the doors and windows open and wind blowing through. The second woman dreamed of a large tree that fell in front of the house. Both the empty house and the fallen tree were symbols of the loss those two women experienced. Look online for interpretations of an empty house or a fallen tree in dreams and there are any number of explanations. But given the context, is there any doubt that these women were processing grief and sense of loss in their dreams?

The South African political prisoner turned president Nelson Mandela had a similar dream after his mother and eldest son died while he was imprisoned on Robben Island. There, he had a recurring dream that he was released from a prison in Johannesburg and walked through the city, which was deserted, arriving at his home in Soweto hours later only to find "a ghost house, with all the doors and windows open, but no one at all there."[5]

Revisiting the common dream related to a final exam in school: Maybe you overslept and missed it, or arrived late, or arrived at the wrong classroom, or studied the wrong material. Maybe you showed up for the test nude, or it's written in a language you don't understand. If you have this dream the night before an actual exam, it is clearly a simple product of your anxiety about the test. But this dream persists for many people well into middle age. Why do we have this dream long after we've left school—and how could dreams like these be not only unsettling but relevant?

Let's return to the two fundamental elements of the dream. The first is the emotion of the dream and the intensity of the emotion; this dream typically provokes intense feelings of anxiety or fear. The second is the central image: an exam in school. Here, it's important to think metaphorically. Unless you are still a student, it's unlikely the dream is about school or tests. Just as the veterans returned to war dreams when they faced marital troubles, anxiety is prompting you to latch onto another time in your life when you were anxious about something.

Harvard psychologist Deirdre Barrett says an exam is a time when someone in a position of authority is evaluating our performance, deciding whether we pass the test—or fail. The image of an exam may be a proxy for something going on in our lives where we are feeling tested or judged. If you have this dream, it makes sense to ask yourself if you're worried you're not meeting someone's expectations.

According to Barrett, school may also be the place where we first experienced other deep feelings, such as embarrassment, stress, and inadequacy. No wonder school and tests serve as a metaphor no matter how old we are. One of the functions of dreams is processing memories and gauging how new experiences fit with old. Dreams about a final exam are likely a way to measure the current anxiety against a past fear that provoked a significant depth of anxiety.

Taking the time to consider the meaning of your dreams requires introspection and self-awareness. Dreams invite us to look deeper into ourselves and examine what they are telling us. Spending time examining the meaning of your dreams can increase both your awareness and your acceptance of your emotions, can lead to important insight about your life, and can lead to a greater sense of wellbeing.

Conclusion

The Transcendent Power of Dreams

In 2016, an 87-year-old man was taken to Vancouver General Hospital after a fall. At the hospital, he began experiencing seizures. His scalp was hooked up to an EEG. By monitoring his brainwaves, physicians hoped to learn more about the seizures. They wound up learning something more profound.

While still attached to the EEG, the man's heart fluttered, then stopped. He had left clear Do Not Resuscitate instructions. With DNR at the top of his chart, nothing was done to restart his heart and revive the patient, and, in his final moments of life, with his heart stopped and the color draining from his body, the EEG continued recording his brain activity. The brainwaves of this dying patient showed something startling.

A long-held assumption among physicians and scientists was that the dying brain would exhibit little activity, or that the activity would quickly dwindle to nothing. This is what happens in other organs. They characteristically whimper into oblivion.

Yet for this man, in the thirty seconds *after* his heart stopped, the brainwaves were fierce, their signals like those seen in both memory recall and dreaming. Other reports are showing similar findings, which raise an intriguing possibility: Death itself may offer one final dream. And that we do not go quietly into the night.

Throughout history, dreams have been seen as the product of supernatural forces, visions delivered by the gods or the spirits to the sleeping mind that reveal something fundamental about ourselves and the world. Ancient cultures were not entirely wrong to think of dreams as supernatural. Indeed, they are a superpower we all share, a unique world each of us illuminates for our own benefit.

Today, we are no different. We, too, sense the power of dreams. Dreams give us an opportunity to evolve and grow. They have the potential to add meaning and richness to our lives, to give us insights about ourselves and others, to reveal what is hidden from us during the day, and to lead us to new paths of understanding and creativity. Dreaming is meaning added, not concealed, to the essential stages of life and the intensely emotional moments that punctuate them.

Dreams drive the emotional centers in our brain to an intensity not possible during our waking life. The Imagination Network is never more active or more free than during our nightly travels. In our everyday life, we often think of our emotional brain as something that can get in the way of making effective decisions or being our most productive selves. In reality, optimal decision-making relies on emotion. We lack the social and situational awareness without it. Patients whose emotional, limbic systems have been damaged struggle to make any decision at all. That means the hyper-emotional experience only possible in our dreams has the potential to provide a unique portal to self-reflection and understanding.

Each night, the brain from which our consciousness and self-awareness arise provides us with a process liberated from the constraints of habit and the limits of our daily existence. This book has aimed to explore not only what we know about the dreaming brain, but most significantly the many ways that our sleeping lives relate to our waking hours. Our dreaming and waking selves are not separate. Understanding how they are intertwined, we can begin to appreciate the power of dreams.

Dreaming gives us the mental capacity to improve the versatility of our thinking, emotions, and instincts. A life spent dreaming expands what we see as possible. In their wildness, dreams give us an important evolutionary advantage, an adaptive mind. This personal genius is built into the system.

Neuroscience has made massive advances in the tools we have to monitor the brain in real time. We can now even record activity at the level of a single neuron. Yet shining the bright light of research on the mysteries of our dreaming mind has not flattened or dulled them. Far from it. The ability to understand dreaming as never before has made it more dazzling, more mysterious—a bit of magic in our quantified world.

Throughout the book, I have tried to explain why and how we dream and the unimaginable complexity that governs us. Even with the most sophisticated, exotic measures of the human brain, I believe we have only taken a glimpse.

In my own life, each day I try to navigate not only the world outside me but the inner world of my mind. The wilder reaches explored in our dreams and dreaming are not distractions to be tamed or ignored. They reveal deeper complexities of consciousness, cognition, and emotion, allowing the whole person to emerge. To ponder the meaning of dreams and dreaming is to explore the meaning of life itself. I believe the astonishing breadth of our dreamscape, from the most frightening conceptions to transcendent revelations, is the ultimate gift of the human mind.

Acknowledgments

Venetia Butterfield for inspiration and a shared vision. Nina Rodríguez-Marty for the masterful editorial scalpel and belief that the book is important. Anna Argenio for moving this idea forward from conception to print and the numerous vital steps that too often go unrecognized. Vanessa Phan for carrying the baton and making the manuscript more. Laurie Ip Fung Chun for her crucial role as managing editor. Alice Dewing and Ania Gordon for presenting this material to the UK and beyond in its most favorable light. Julia Falkner made it a priority for US media to see its potential and her success has elevated the book's reach. Raven Ross attentively drove marketing in the US. Amelia Evans, Monique Corless, and the rest of the Penguin rights team for enticing the world with its value. Richard Kilgariff for adding momentum. David Steen Martin for its shared construction.

In exploring the depths of what we dream and how we dream, it must be recognized that the reports, publications, and science that comprise our knowledge to this point have not included the breadth of our humanity, and important stories have yet to be heard. I look forward to the scientific community incorporating the valuable insights gained as more diverse voices are included—which will deepen the nuance, and therefore understanding, of why we dream. As with my dreams, this book is 100 percent human generated.

Notes

Introduction: A Nightly Dose of Wonder

1 Byron, George Gordon, Lord, "The Dream," public-domain-poetry.com / george-gordon-byron / dream-10617

1. *We Have Evolved to Dream*

1 Pace-Schott, Edward F., "Dreaming as a Storytelling Instinct," *Frontiers in Psychology*, April 2, 2013

2 Hall, Calvin S., and Van de Castle, Robert L.,' *The Content Analysis of Dreams*, Appleton-Century-Crofts, 1966

3 Domhoff, William, and Schneider, Adam, "Are Dreams Social Simulations? Or Are They Enactments of Conceptions and Personal Concerns? An Empirical and Theoretical Comparison of Two Dream Theories," *Dreaming*, 2018

4 Bowe-Anders, Constance, et al., "Effects of Goggle-altered Color Perception on Sleep," *Perceptual and Motor Skills*, February 1974

5 De Koninck, Joseph et al., "Vertical Inversion of the Visual Field and REM Sleep Mentation," *Journal of Sleep Research*, March 1996

6 Arnulf, Isabelle, et al., "Will Students Pass a Competitive Exam That They Failed in Their Dreams?," *Consciousness and Cognition*, October 2014

7 van der Helm, Els, et al., "REM Sleep Depotentiates Amygdala Activity to Previous Emotional Experiences," *Current Biology*, December 6, 2011

8 Cartwright, Rosalind, et al., "Broken Dreams: A Study of the Effects of Divorce and Depression on Dream Content," *Psychiatry*, 1984

9 Flinn, Mark V., "The Creative Neurons," *Frontiers in Psychology*, November 22, 2021

10 Hoel, Erik, "The Overfitted Brain: Dreams Exist to Assist Generalization," *Patterns*, May 14, 2021

2. *We Need Nightmares*

1 "Nightmare on Science Street," *Science Vs* podcast, June 9, 2022

2 Elder, Rachel, "Speaking Secrets: Epilepsy, Neurosurgery, and Patient Testimony in the Age of the Explorable Brain, 1934–1960," *Bulletin of the History of Medicine*, Winter 2015

3 Hublin, Christer, et al., "Nightmares: Familial Aggregation and Association with Psychiatric Disorders in a Nationwide Twin Cohort," *American Journal of Medical Genetics*, October 25, 2002

4 Moore, Rebecca S., et al., "Piwi/PRG-1 Argonaute and TGF-β Mediate Transgenerational Learned Pathogenic Avoidance," *Cell*, June 13, 2019

5 Arzy, Shahar, et al., "Induction of an Illusory Shadow Person," *Nature*, September 2006

6 Krakow, Barry, et al., "Imagery Rehearsal Therapy for Chronic Nightmares in Sexual Assault Survivors with Posttraumatic Stress Disorder: A Randomized Controlled Trial," *Journal of the American Medical Association*, August 1, 2001

3. Erotic Dreams: The Embodiment of Desire

1 Quiroga, Rodrigo Quian, "Single-neuron Recordings in Epileptic Patients," *Advances in Clinical Neuroscience and Rehabilitation*, July / August 2009

2 DreamBank.net, a searchable collection of more than 20,000 dream reports

3 Chen, Wanzhen, et al., "Development of a Structure-validated Sexual Dream Experience Questionnaire (SDEQ) in Chinese University Students," *Comprehensive Psychiatry*, January 2015

4 Selterman, Dylan F., et al., "Dreaming of You: Behavior and Emotion in Dreams of Significant Others Predict Subsequent Relational Behavior," *Social Psychological and Personality Science*, May 6, 2013

5 Domhoff, G. William, "Barb Sanders: Our Best Case Study to Date, and One That Can Be Built Upon," dreams.ucsc.edu/Findings/barb_sanders.html, undated

4. Dreaming and Creativity: How Dreams Unlock the Creative Within

1 Dement, William, *Some Must Watch While Some Must Sleep*, W. H. Freemont & Co., 1972, pp. 99–101

2 Liu, Siyuan, et al., "Brain Activity and Connectivity during Poetry Composition: Toward a Multidimensional Model of the Creative Process," *Human Brain Mapping*, May 26, 2015

3 Cai, Denise J., et al., "REM, Not Incubation, Improves Creativity by Priming Associative Networks," *Proceedings of the National Academy of Sciences*, June 23, 2009

4 Mason, Robert A., and Just, Marcel Adam, "Neural Representations of Procedural Knowledge," *Psychological Science*, May 12, 2020

5 Hartmann, Ernest, et al., "Who Has Nightmares? The Personality of the Lifelong Nightmare Sufferer," *Archives of General Psychiatry*, January 1987

6 Barrett, Deirdre, "Dreams and Creative Problem-solving," *Annals of the New York Academy of Sciences*, June 22, 2017

7 "BAFTA Screenwriters' Lecture Series," September 30, 2011, youtube.com

8 Dalí, Salvador, *50 Secrets of Magic Craftsmanship* (transl. by H. Chevalier), Dover, 1992

9 Lacaux, Célia, et al., "Sleep Onset Is a Creative Sweet Spot," *Science Advances*, December 8, 2021

10 Horowitz, Adam Haar, et al., "Dormio: A Targeted Dream Incubation Device," *Consciousness and Cognition*, August 2020

5. Dreaming and Health: What Dreams Reveal About Our Wellbeing

1 Kasatkin, Vasily, *A Theory of Dreams*, lulu.com, May 27, 2014

2 Rozen, Naama, and Soffer-Dudek, Nirit, "Dreams of Teeth Falling Out: An Empirical Investigation of Physiological and Psychological Correlates," *Frontiers in Psychology*, September 26, 2018

3 Cartwright, Rosalind, "Dreams and Adaptation to Divorce," in *Trauma and Dreams*, ed. Deirdre Barrett, Harvard University Press, 1996, pp. 179–185

4 Hill, Clara, and Knox, Sarah, "The Use of Dreams in Modern Psychotherapy," *International Review of Neurobiology*, 2010

5 Duffey, Thelma H., et al., "The Effects of Dream Sharing on Marital Intimacy and Satisfaction," *Journal of Couple & Relationship Therapy*, September 25, 2008

6 DeHart, Dana, "Cognitive Restructuring Through Dreams and Imagery: Descriptive Analysis of a Women's Prison-based Program," *Journal of Offender Rehabilitation*, December 22, 2009
7 Blagrove, Mark, et al., "Testing the Empathy Theory of Dreaming: The Relationships between Dream Sharing and Trait and State Empathy," *Frontiers in Psychology*, June 20, 2019
8 Ullman, Montague, "The Experiential Dream Group: Its Application in the Training of Therapists," *Dreaming*, December 1994
9 Cartwright, Rosalind, et al., "REM Sleep Reduction, Mood Regulation and Remission in Untreated Depression," *Psychiatry Research*, December 1, 2003
10 da Silva, Thiago Rovai, and Nappo, Solange Aparecida, "Crack Cocaine and Dreams: The View of Users," *Ciencia & Saude Coletiva*, March 24, 2019
11 "The Dreaming Mind: Waking the Mysteries of Sleep," World Science Festival, November 17, 2022, youtube.com
12 van der Kolk, Bessel, *The Body Keeps the Score: Brain, Mind, and Body in the Healing of Trauma*, Viking, 2014
13 Hartmann, Ernest, "Nightmare after Trauma as Paradigm for All Dreams: A New Approach to the Nature and Functions of Dreaming," *Psychiatry: Interpersonal and Biological Processes*, 1998
14 Li, Hao, et al., "Neurotensin Orchestrates Valence Assignment in the Amygdala," *Nature*, August 18, 2022

6. Lucid Dreams: A Hybrid of the Waking and Dreaming Minds

1 Hearne, Keith M. T., "Lucid Dreams: An Electro-physiological and Psychological Study," doctoral thesis, University of Liverpool, May 1978

2 Worsley, Alan, "Alan Worsley's Work on Lucid Dreaming," *Lucidity Letter*, 1991

3 Hearne, Keith M. T., *The Dream Machine: Lucid Dreams and How to Control Them*, Aquarian Press, 1990

4 Mallett, Remington, "Partial Memory Reinstatement while (Lucid) Dreaming to Change the Dream Environment," *Consciousness and Cognition*, 2020

5 LaBerge, Stephen, "Lucid Dreaming and the Yoga of the Dream State: A Psychophysiological Perspective," in *Buddhism and Science: Breaking New Ground*, ed. B. A. Wallace, Columbia University Press, 2003, p. 233

6 "Lucid Dreaming with Ursula Voss," Science & Cocktails, youtube.com

7 Zhunusova, Zanna, Raduga, Michael, and Shashkov, Andrey, "Overcoming Phobias by Lucid Dreaming," *Psychology of Consciousness: Theory, Research, and Practice*, 2022

8 Erlacher, Daniel, Stumbrys, Tadas, and Schredl, Michael, "Frequency of Lucid Dreams and Lucid Dream Practice in German Athletes," *Imagination, Cognition and Personality*, February 2012

9 Schädlich, Melanie, Erlacher, Daniel, and Schredl, Michael, "Improvement of Darts Performance following Lucid Dream Practice Depends on the Number of Distractions while Rehearsing within the Dream—A Sleep Laboratory Pilot Study," *Journal of Sports Sciences*, December 22, 2016

10 Schädlich, Melanie, and Erlacher, Daniel, "Lucid Music—A Pilot Study Exploring the Experiences and Potential of Music-making in Lucid Dreams," *Dreaming*, 2018

11 "The Dreaming Mind: Waking the Mysteries of Sleep," World Science Festival, youtube.com

12 Stumbrys, Tadas, and Daniels, Michael, "An Exploratory Study of Creative Problem Solving in Lucid Dreams: Preliminary

Findings and Methodological Considerations," *International Journal of Dream Research*, November 2010

13 "The Dreaming Mind: Waking the Mysteries of Sleep," World Science Festival, youtube.com

14 Konkoly, Karen R., et al., "Real-time Dialogue between Experimenters and Dreamers during REM Sleep," *Current Biology*, April 12, 2021

15 Raduga, Michael, "'I Love You': The First Phrase Detected from Dreams," *Sleep Science*, 2022

7. How to Induce Lucid Dreams

1 Erlacher, Daniel, Stumbrys, Tadas, and Schredl, Michael, "Frequency of Lucid Dreams and Lucid Dream Practice in German Athletes," *Imagination, Cognition and Personality*, February 2012

2 Cosmic Iron, "Senses Initiated Lucid Dream (SSILD) Official Tutorial," cosmiciron.blogspot.com/2013/01/senses-initiated-lucid-dream-ssild_16.html

3 Appel, Kristoffer, "Inducing Signal-verified Lucid Dreams in 40% of Untrained Novice Lucid Dreamers within Two Nights in a Sleep Laboratory Setting," *Consciousness and Cognition*, August 2020

4 LaBerge, Stephen, LaMarca, Kristen, and Baird, Benjamin, "Pre-sleep Treatment with Galantamine Stimulates Lucid Dreaming: A Double-blind, Placebo-controlled, Crossover Study," *PLOS One*, 2018

5 LaBerge, Stephen, and Levitan, Lynn, "Validity Established of DreamLight Cues for Eliciting Lucid Dreaming," *Dreaming*, 1995

6 Mota-Rolim, Sérgio A. et al., "Portable Devices to Induce Lucid Dreams—Are They Reliable?," *Frontiers in Neuroscience*, May 8, 2019

8. *The Future of Dreaming*

1 "Yukiyasu Kamitani (Kyoto University), Deep Image Reconstruction from the Human Brain," youtube.com
2 Huth, Alexander G., et al., "Natural Speech Reveals the Semantic Maps That Tile Human Cerebral Cortex," *Nature*, April 27, 2016
3 Popham, Sarah F., et al., "Visual and Linguistic Semantic Representations Are Aligned at the Border of Human Visual Cortex," *Nature Neuroscience*, November 2021
4 Shanahan, Laura K., et al., "Odor-evoked Category Reactivation in Human Ventromedial Prefrontal Cortex during Sleep Promotes Memory Consolidation," *Neuroscience*, December 18, 2018
5 Arzi, Anat, et al., "Olfactory Aversive Condition during Sleep Reduces Cigarette-smoking Behavior," *The Journal of Neuroscience*, November 12, 2014
6 Mahdavi, Mehdi, Fatehi-Rad, Navid, and Barbosa, Belem, "The Role of Dreams of Ads in Purchase Intention," *Dreaming*, 2019
7 Ai, Sizhi, et al., "Promoting Subjective Preferences in Simple Economic Choices during Nap," *eLife*, December 6, 2018
8 *The Risks and Challenges of Neurotechnologies for Human Rights*, UNESCO, 2023
9 "Rafael Yuste: 'Let's Act Before It's Too Late'," en.unesco.org/courier/2022-1/rafael-yuste-lets-act-its-too-late, 2022

9. *The Interpretation of Dreams*

1 Malinowski, Josie, and Horton, C. L., "Dreams Reflect Nocturnal Cognitive Processes: Early-night Dreams Are More Continuous with Waking Life, and Late-night Dreams Are More Emotional and Hyperassociative," *Consciousness and Cognition*, 2021

2 Hartmann, Ernest, "The Underlying Emotion and the Dream: Relating Dream Imagery to the Dreamer's Underlying Emotion Can Help Elucidate the Nature of Dreaming," *International Review of Neurobiology*, 2010

3 Breger, L., Hunter, I., and Lane, R., "The Effect of Stress on Dreams," *Psychological Issues*, 1971

4 Hartmann, Ernest, "The Underlying Emotion and the Dream: Relating Dream Imagery to the Dreamer's Underlying Emotion Can Help Elucidate the Nature of Dreaming," *International Review of Neurobiology*, 2010

5 Truscott, Ross, "Mandela's Dreams," africasacountry.com/2018/11/mandelas-dreams, November 15, 2018

Further Reading

Ahmadi, Fereshteh, and Hussin, Nur Atikah Mohamed, "Cancer Patients' Meaning Making Regarding Their Dreams: A Study among Cancer Patients in Malaysia," *Dreaming*, 2020

Akkaoui, Marine Ambar, et al., "Nightmares in Patients with Major Depressive Disorder, Bipolar Disorder, and Psychotic Disorders: A Systematic Review," *Journal of Clinical Medicine*, 2020

Alcaro, Antonio, and Carta, Stefano, "The 'Instinct' of Imagination: A Neuro-ethological Approach to the Evolution of the Reflective Mind and Its Application to Psychotherapy," *Frontiers in Human Neuroscience*, January 23, 2019

Alessandria, Maria, et al., "Normal Body Scheme and Absent Phantom Limb Experience in Amputees while Dreaming," *Consciousness and Cognition*, July 13, 2011

Alexander, Marcalee Sipski, and Marson, Lesley, "The Neurologic Control of Arousal and Orgasm with Specific Attention to Spinal Cord Lesions: Integrating Preclinical and Clinical Sciences," *Autonomic Neuroscience: Basic and Clinical*, 2018

Andersen, Monica L., et al., "Sexsomnia: Abnormal Sexual Behavior during Sleep," *Brain Research Reviews*, 2007

Andrews-Hanna, Jessica R., "The Brain's Default Network and Its Adaptive Role in Internal Mentation," *Neuroscientist*, June 2012

Andrews-Hanna, Jessica R., and Grilli, Matthew D., "Mapping the Imaginative Mind: Charting New Paths Forward," *Current Directions in Psychological Science*, February 2021

Appel, K., et al., "Inducing Signal-verified Lucid Dreams in 40% of Untrained Novice Lucid Dreamers within Two Nights in a Sleep Laboratory Setting," *Consciousness and Cognition*, 2020

Arehart-Treichel, Joan, "Amazon People's Dreams Hold Lessons for Psychotherapy," *Psychiatric News*, March 4, 2011

Aspy, Denholm J., "Findings from the International Dream Induction Study," *Frontiers in Psychology*, July 17, 2020

Aspy, Denholm J., et al., "Reality Testing and the Mnemonic Induction of Lucid Dreams: Findings from the National Australian Lucid Dream Induction Study," *Dreaming*, 2017

BaHammam, Ahmed S., and Almeneessier, Aljohara S., "Dreams and Nightmares in Patients with Obstructive Sleep Apnea: A Review," *Frontiers in Neurology*, October 22, 2019

Bainbridge, Wilma A., et al., "Quantifying Aphantasia Through Drawing: Those without Visual Imagery Show Deficits in Object but Not Spatial Memory," *Cortex*, 2021

Baird, Benjamin, et al., "Frequent Lucid Dreaming Associated with Increased Functional Connectivity between Frontopolar Cortex and Temporoparietal Association Areas," *Scientific Reports*, December 12, 2018

Baird, Benjamin, et al., "Inspired by Distraction: Mind Wandering Facilitates Creative Incubation," *Psychological Science*, October 1, 2012

Baird, Benjamin, LaBerge, Stephen, and Tononi, Giulio, "Two-way Communication in Lucid REM Sleep Dreaming," *Trends in Cognitive Sciences*, June 2021

Baird, Benjamin, Mota-Rolim, Sergio, and Dresler, Martin, "The Cognitive Neuroscience of Lucid Dreaming," *Neuroscience Biobehavioral Review*, May 1, 2020

Baird, Benjamin, Tononi, Giulio, and LaBerge, Stephen, "Lucid Dreaming Occurs in Activated Rapid Eye Movement Sleep, Not a Mixture of Sleep and Wakefulness," *Sleep*, 2022

Balasubramaniam, B., and Park, G. R., "Sexual Hallucinations during and after Sedation and Anaesthesia," *Anaesthesia*, 2003

Baldelli, Luca, and Provini, Federica, "Differentiating Oneiric Stupor in Agrypnia Excitata from Dreaming Disorders," *Frontiers in Neurology*, November 12, 2020

Ball, Tonio, et al., "Signal Quality of Simultaneously Recorded Invasive and Non-invasive EEG," *NeuroImage*, 2009

Barnes, Christopher M., Watkins, Trevor, and Klotz, Anthony, "An Exploration of Employee Dreams: The Dream-based Overnight Carryover of Emotional Experiences at Work," *Sleep Health*, 2021

Barrett, Deirdre, "Dreams about COVID-19 versus Normative Dreams: Trends by Gender," *Dreaming*, 2020

Barrett, Deirdre, "Dreams and Creative Problem-solving," *Annals of the New York Academy of Sciences*, June 22, 2017

Barrett, Deirdre, "The 'Committee of Sleep': A Study of Dream Incubation for Problem Solving," *Dreaming*, 1993

Barrett, Deirdre, "The Dream Character as Prototype for Multiple Personality Alter," *Dissociation*, March 1995

Barry, Daniel N., et al., "The Neural Dynamics of Novel Scene Imagery," *The Journal of Neuroscience*, May 29, 2019

Bashford, Luke, et al., "The Neurophysiological Representation of Imagined Somatosensory Percepts in Human Cortex," *The Journal of Neuroscience*, March 10, 2021

Bastin, Julien, et al., "Direct Recordings from Human Anterior Insula Reveal Its Leading Role within the Error-monitoring Network," *Cerebral Cortex*, February 2017

Baylor, George W., and Cavallero, Corrado, "Memory Sources Associated with REM and NREM Dream Reports Throughout the Night: A New Look at the Data," *Sleep*, 2001

Beaty, Roger E., et al., "Brain Networks of the Imaginative Mind: Dynamic Functional Connectivity of Default and Cognitive Control Networks Relates to Openness to Experience," *Human Brain Mapping*, 2017

Beaty, Roger E., et al., "Creative Constraints: Brain Activity and Network Dynamics Underlying Semantic Interference during Idea Production," *NeuroImage*, 2017

Beaty, Roger E., et al., "Creativity and the Default Network: A Functional Connectivity Analysis of the Creative Brain at Rest," *Neuropsychologia*, 2014

Beaty, Roger E., et al., "Personality and Complex Brain Networks: The Role of Openness to Experience in Default Network Efficiency," *Human Brain Mapping*, 2016

Beaty, Roger E., Silvia, Paul J., and Benedek, Mathias, "Brain Networks Underlying Novel Metaphor Production," *Brain and Cognition*, 2017

Beck, Jane C., "'Dream Messages' from the Dead," *Journal of the Folklore Institute*, December 1973

Bekrater-Bodmann, Robin, et al., "Post-amputation Pain Is Associated with the Recall of an Impaired Body Representation in Dreams—Results from a Nation-wide Survey on Limb Amputees," *PLOS One*, March 5, 2015

Belinda, Casher D., and Christian, Michael S., "A Spillover Model of Dreams and Work Behavior: How Dream Meaning Ascription Promotes Awe and Employee Resilience," *Academy of Management*, June 27, 2022

Beversdorf, David Q., "Neuropsychopharmacological Regulation of Performance on Creativity-related Tasks," *Current Opinion in Behavioral Sciences*, 2019

Bhat, Sushanth, et al., "Dream-enacting Behavior in Non-rapid Eye Movement Sleep," *Sleep Medicine*, 2012

Blagrove, Mark, Farmer, Laura, and Williams, Elvira, "The Relationship of Nightmare Frequency and Nightmare Distress to Well-being," *Journal of Sleep Research*, 2004

Blagrove, Mark, and Pace-Schott, Edward F., "Trait and Neurobiological Correlates of Individual Differences in Dream Recall and Dream Content," *International Review of Neurobiology*, 2010

Blanchette-Carrière, Cloé, et al., "Attempted Induction of Signalled Lucid Dreaming by Transcranial Alternating Current Stimulation," *Consciousness and Cognition*, 2020

Błaśkiewicz, Monika, "Healing Dreams at Epidaurus: Analysis and Interpretation of the Epidaurian Iamata," *Miscellanea Anthropologica et Sociologica*, 2014

Boehme, Rebecca, and Olausson, Håkan, "Differentiating Self-touch from Social Touch," *Current Opinion in Behavioral Sciences*, 2022

Bogzaran, Fariba, "Experiencing the Divine in the Lucid Dream State," *Lucidity Letter*, 1991

Bonamino, C., Watling, C., and Polman, R., "The Effectiveness of Lucid Dreaming Practice on Waking Task Performance: A Scoping Review of Evidence and Meta-analysis," *Dreaming*, 2022

Borchers, Svenja, et al., "Direct Electrical Stimulation of Human Cortex—the Gold Standard for Mapping Brain Functions?" *Nature Reviews Neuroscience*, November 2011

Borghi, Lidia, et al., "Dreaming during Lockdown: A Quali-quantitative Analysis of the Italian Population Dreams during the First COVID-19 Pandemic Wave," *Research in Psychotherapy: Psychopathology, Process and Outcome*, 2021

Bradley, Claire, et al., "State-dependent Effects of Neural Stimulation on Brain Function and Cognition," *Nature Reviews Neuroscience*, August 2022

Braun, A. R., et al., "Regional Cerebral Blood Flow Throughout the Sleep–Wake Cycle: An H2(15)O PET Study," *Brain*, 1997

Brecht, Michael, Lenschow, Constanze, and Rao, Rajnish P., "Socio-sexual Processing in Cortical Circuits," *Current Opinion in Neurobiology*, 2018

Brink, Susan M., Allan, John A. B., and Boldt, Walter, "Symbolic Representation of Psychological States in the Dreams of Women with Eating Disorders," *Canadian Journal of Counselling/Revue canadienne de counseling*, 1995

Brock, Matthew S., et al., "Clinical and Polysomnographic Features of Trauma-associated Sleep Disorder," *Journal of Clinical Sleep Medicine*, 2022

Brosch, Renate, "What We 'See' When We Read: Visualization and Vividness in Reading Fictional Narratives," *Cortex*, 2018

Brugger, Peter, "The Phantom Limb in Dreams," *Consciousness and Cognition*, 2008

Bugalho, Paulo, et al., "Progression in Parkinson's Disease: Variation in Motor and Non-motor Symptoms Severity and Predictors of Decline in Cognition, Motor Function, Disability, and Health-related Quality of Life as Assessed by Two Different Methods," *Movement Disorders Clinical Practice*, June 2021

Bugalho, Paulo, and Paiva, Teresa, "Dream Features in the Early Stages of Parkinson's Disease," *Journal of Neural Transmission*, 2011

Bulgarelli, Chiara, et al., "The Developmental Trajectory of Fronto-temporoparietal Connectivity as a Proxy of the Default Mode Network: A Longitudinal fNIRS Investigation," *Human Brain Mapping*, March 4, 2020

Bulkeley, Kelly, "Dreaming as Inspiration: Evidence from Religion, Philosophy, Literature, and Film," *International Review of Neurobiology*, 2010

Bulkeley, Kelly, "The Future of Dream Science," *Annals of the New York Academy of Sciences*, 2017

Burk, Larry, "Warning Dreams Preceding the Diagnosis of Breast Cancer: A Survey of the Most Important Characteristics," *Explore*, June 2015

Burnham, Melissa M., and Conte, Christian, "Developmental Perspective Dreaming across the Lifespan and What This Tells Us," *International Review of Neurobiology*, 2010

Bushnell, Greta A., et al., "Association of Benzodiazepine Treatment for Sleep Disorders with Drug Overdose Risk among Young People," *JAMA Network Open*, 2022

Calabrò, Rocco S., et al., "Neuroanatomy and Function of Human Sexual Behavior: A Neglected or Unknown Issue?" *Brain and Behavior*, 2019

Campbell, Ian G., et al., "Sex, Puberty, and the Timing of Sleep EEG Measured Adolescent Brain Maturation," *Proceedings of the National Academy of Sciences*, March 26, 2012

Cappadona, R., et al., "Sleep, Dreams, Nightmares, and Sex-related Differences: A Narrative Review," *European Review for Medical and Pharmacological Sciences*, 2021

Carr, Michelle, et al., "Dream Engineering: Simulating Worlds Through Sensory Stimulation," *Consciousness and Cognition*, 2020

Carr, Michelle, et al., "Towards Engineering Dreams," *Consciousness and Cognition*, 2020

Carton-Leclercq, Antoine, et al., "Laminar Organization of Neocortical Activities during Systemic Anoxia," *Neurobiology of Disease*, November 2023

Cartwright, Rosalind, et al., "Effect of an Erotic Movie on the Sleep and Dreams of Young Men," *Archives of General Psychiatry*, March 1969

Cartwright, Rosalind, et al., "Relation of Dreams to Waking Concerns," *Psychiatry Research*, 2006

Carvalho, Diana, et al., "The Mirror Neuron System in Post-stroke Rehabilitation," *International Archives of Medicine*, 2013

Carvalho, I., et al., "Cultural Explanations of Sleep Paralysis: The Spiritual Phenomena," *European Psychiatry*, March 23, 2020

Cavallero, Corrado, "The Quest for Dream Sources," *Journal of Sleep Research*, 1993

Cavallotti, Simone, et al., "Aggressiveness in the Dreams of Drug-naïve and Clonazepam-treated Patients with Isolated REM Sleep Behavior Disorder," *Sleep Medicine*, March 5, 2022

Chaieb, Leila, et al., "New Perspectives for the Modulation of Mind-wandering Using Transcranial Electric Brain Stimulation," *Neuroscience*, 2019

Chellappa, Sarah Laxhmi, and Cajochen, Christian, "Ultradian and Circadian Modulation of Dream Recall: EEG Correlates and Age Effects," *International Journal of Psychophysiology*, 2013

Childress, Anna Rose, et al., "Prelude to Passion: Limbic Activation by 'Unseen' Drug and Sexual Cues," *PLOS One*, January 2008

Choi, S. Y., "Dreams as a Prognostic Factor in Alcoholism," *The American Journal of Psychiatry*, 1973

Christo, George, and Franey, Christine, "Addicts Drug-related Dreams: Their Frequency and Relationship to Six-month Outcomes," *Substance Use & Misuse*, 1996

Christoff, Kalina, et al., "Mind-wandering as Spontaneous Thought: A Dynamic Framework," *Nature Reviews Neuroscience*, November 2016

Cicolin, Alessandro, et al., "End-of-life in Oncologic Patients' Dream Content," *Brain Sciences*, August 1, 2020

Cinosi, E., et al., "Sleep Disturbances in Eating Disorders: A Review," *La Clinica Terapeutica*, November 2011

Cipolli, Carlo, et al., "Beyond the Neuropsychology of Dreaming: Insights into the Neural Basis of Dreaming with New Techniques of Sleep Recording and Analysis," *Sleep Medicine Reviews*, 2017

Clarke, Jessica, DeCicco, Teresa L., and Navara, Geoff, "An Investigation among Dreams with Sexual Imagery, Romantic Jealousy and Relationship Satisfaction," *International Journal of Dream Research*, 2010

Cochen, V., et al., "Vivid Dreams, Hallucinations, Psychosis and REM Sleep in Guillain–Barré Syndrome," *Brain*, 2005

Colace, Claudio, "Drug Dreams in Cocaine Addiction," *Drug and Alcohol Review*, March 2006

Collerton, Daniel, and Perry, Elaine, "Dreaming and Hallucinations—Continuity or Discontinuity? Perspectives from Dementia with Lewy Bodies," *Consciousness and Cognition*, 2011

Conte, Francesca, et al., "Changes in Dream Features across the First and Second Waves of the Covid-19 Pandemic," *Journal of Sleep Research*, June 22, 2021

Coolidge, Frederick L., et al., "Do Nightmares and Generalized Anxiety Disorder in Childhood and Adolescence Have a Common Genetic Origin?" *Behavior Genetics*, November 10, 2009

Cooper, Shelly, "Lighting up the Brain with Songs and Stories," *General Music Today*, 2010

Courtois, Frédérique, Alexander, Marcalee, and McLain, Amie B. Jackson, "Women's Sexual Health and Reproductive Function after SCI," *Topics in Spinal Cord Injury Rehabilitation*, 2017

Coutts, Richard, "Variation in the Frequency of Relationship Characters in the Dream Reports of Singles: A Survey of 15,657 Visitors to an Online Dating Website," *Comprehensive Psychology*, 2015

Cox, Ann, "Sleep Paralysis and Folklore," *Journal of the Royal Society of Medicine Open*, 2015

Curot, Jonathan, et al., "Déjà-rêvé: Prior Dreams Induced by Direct Electrical Brain Stimulation," *Brain Stimulation*, 2018

Curot, Jonathan, et al., "Memory Scrutinized Through Electrical Brain Stimulation: A Review of 80 Years of Experiential Phenomena," *Neuroscience and Biobehavioral Reviews*, 2017

Dagher, Alain, and Misic, Bratislav, "Holding onto Youth," *Cell Metabolism*, August 1, 2017

Dahan, Lionel, et al., "Prominent Burst Firing of Dopaminergic Neurons in the Ventral Tegmental Area during Paradoxical Sleep," *Neuropsychopharmacology*, 2007

Dale, Allyson, Lafrenière, Alexandre, and De Koninck, Joseph, "Dream Content of Canadian Males from Adolescence to Old Age: An Exploration of Ontogenetic Patterns," *Consciousness and Cognition*, March 2017

Dale, Allyson, Lortie-Lussier, Monique, and De Koninck, Joseph, "Ontogenetic Patterns in the Dreams of Women across the Lifespan," *Consciousness and Cognition*, 2015

Dang-Vu, T. T., et al., "A Role for Sleep in Brain Plasticity," *Journal of Pediatric Rehabilitation Medicine*, 2006

D'Argembeau, Arnaud, and Van der Linden, Martial, "Individual Differences in the Phenomenology of Mental Time Travel: The Effect of Vivid Visual Imagery and Emotion Regulation Strategies," *Consciousness and Cognition*, 2006

Davis, Joanne L., and Wright, David C., "Case Series Utilizing Exposure, Relaxation, and Rescripting Therapy: Impact on Nightmares, Sleep Quality, and Psychological Distress," *Behavioral Sleep Medicine*, 2005

Dawes, Alexei J., et al., "A Cognitive Profile of Multi-sensory Imagery, Memory and Dreaming in Aphantasia," *Scientific Reports*, 2020

DeCicco, Teresa L., et al., "A Cultural Comparison of Dream Content, Mood and Waking Day Anxiety between Italians and Canadians," *International Journal of Dream Research*, 2013

DeCicco, Teresa L., et al., "Exploring the Dreams of Women with Breast Cancer: Content and Meaning of Dreams," *International Journal of Dream Research*, November 2010

De Gennaro, Luigi, et al., "How We Remember the Stuff That Dreams Are Made of: Neurobiological Approaches to the Brain Mechanisms of Dream Recall," *Behavioural Brain Research*, 2012

De Gennaro, Luigi, et al., "Recovery Sleep after Sleep Deprivation Almost Completely Abolishes Dream Recall," *Behavioural Brain Research*, 2010 de la Chapelle, Aurélien, et al., "Relationship between Epilepsy and Dreaming: Current Knowledge, Hypotheses, and Perspectives," *Frontiers in Neuroscience*, September 6, 2021

de Macêdo, Tainá Carla Freitas, et al., "My Dream, My Rules: Can Lucid Dreaming Treat Nightmares?" *Frontiers in Psychology*, November 2019

Dement, William C., "History of Sleep Medicine," *Neurologic Clinics*, 2005

Dement, William C., "The Effect of Dream Deprivation: The Need for a Certain Amount of Dreaming Each Night Is Suggested by Recent Experiments," *Science*, 1960

Denis, Dan, and Poerio, Giulia L., "Terror and Bliss? Commonalities and Distinctions between Sleep Paralysis, Lucid Dreaming, and Their Associations with Waking Life Experiences," *Journal of Sleep Research*, 2017

Desseilles, Martin, et al., "Cognitive and Emotional Processes during Dreaming: A Neuroimaging View," *Consciousness and Cognition*, 2011

Devine, Rory T., and Hughes, Claire, "Silent Films and Strange Stories: Theory of Mind, Gender, and Social Experiences in Middle Childhood," *Child Development*, November 30, 2012

Dijkstra, Nadine, Bosch, Sander E., and van Gerven, Marcel A. J., "Shared Neural Mechanisms of Visual Perception and Imagery," *Trends in Cognitive Sciences*, 2019

Di Noto, Paula M., et al., "The Hermunculus: What Is Known about the Representation of the Female Body in the Brain?" *Cerebral Cortex*, May 2013

Dodet, Pauline, et al., "Lucid Dreaming in Narcolepsy," *Sleep*, 2015

Domhoff, William G., and Schneider, Adam, "From Adolescence to Young Adulthood in Two Dream Series: The Consistency and Continuity of Characters and Major Personal Interests," *Dreaming*, 2020

Domhoff, William G., and Schneider, Adam, "Similarities and Differences in Dream Content at the Cross-cultural, Gender, and Individual Levels," *Consciousness and Cognition*, 2008

Duffau, Hugues, "The 'Frontal Syndrome' Revisited: Lessons from Electrostimulation Mapping Studies," *Cortex*, 2012

Duffey, Thelma H., et al., "The Effects of Dream Sharing on Marital Intimacy and Satisfaction," *Journal of Couples & Relationship Therapy*, 2004

Dumontheil, Iroise, Apperly, Ian A., and Blakemore, Sarah-Jayne, "Online Usage of Theory of Mind Continues to Develop in Late Adolescence," *Developmental Science*, 2010

Dumser, Britta, et al., "Symptom Dynamics among Nightmare Sufferers: An Intensive Longitudinal Study," *Journal of Sleep Research*, October 17, 2022

Durantin, Gautier, Dehais, Frederic, and Delorme, Arnaud, "Characterization of Mind Wandering Using fNIRS," *Frontiers in Systems Neuroscience*, March 26, 2015

Edwards, Christopher L., et al., "Dreaming and Insight," *Frontiers in Psychology*, December 24, 2013

Eichenbaum, Howard, "Time Cells in the Hippocampus: A New Dimension for Mapping Memories," *Nature Reviews Neuroscience*, November 2014

Eickhoff, Simon B., et al., "Anatomical and Functional Connectivity of Cytoarchitectonic Areas within the Human Parietal Operculum," *The Journal of Neuroscience*, May 5, 2010

El Haj, Mohamad, and Lenoble, Quentin, "Eying the Future: Eye Movement in Past and Future Thinking," *Cortex*, 2018

Engel, Andreas K., et al., "Invasive Recordings from the Human Brain: Clinical Insights and Beyond," *Nature Reviews Neuroscience*, January 2005

Erlacher, Daniel, and Chapin, Heather, "Lucid Dreaming: Neural Virtual Reality as a Mechanism for Performance Enhancement," *International Journal of Dream Research*, 2010

Erlacher, Daniel, and Shredl, Michael, "Dreams Reflecting Waking Sports Activities: A Comparison of Sport and Psychology Students," *International Journal of Sport Psychology*, 2004

Erlacher, Daniel, and Shredl, Michael, "Do REM (Lucid) Dreamed and Executed Actions Share the Same Neural Substrate?" *International Journal of Dream Research*, 2008

Erlacher, Daniel, and Shredl, Michael, "Practicing a Motor Task in a Lucid Dream Enhances Subsequent Performance: A Pilot Study," *The Sport Psychologist*, 2010

Erlacher, Daniel, and Shredl, Michael, "Time Required for Motor Activity in Lucid Dreams," *Perceptual and Motor Skills*, 2004

Erlacher, Daniel, Ehrlenspiel, Felix, and Schredl, Michael, "Frequency of Nightmares and Gender Significantly Predict Distressing Dreams of German Athletes Before Competitions or Games," *The Journal of Psychology*, 2011

Erlacher, Daniel, et al., "Inducing Lucid Dreams by Olfactory-cued Reactivation of Reality Testing during Early-morning Sleep: A Proof of Concept," *Consciousness and Cognition*, 2020

Erlacher, Daniel, et al., "Ring, Ring, Ring . . . Are You Dreaming? Combining Acoustic Stimulation and Reality Testing for Lucid Dream Induction: A Sleep Laboratory Study," *International Journal of Dream Research*, 2020

Erlacher, Daniel, et al., "Time for Actions in Lucid Dreams: Effects of Task Modality, Length, and Complexity," *Frontiers in Psychology*, 2014

Erlacher, Daniel, Shredl, Michael, and Stumbrys, Tadas, "Self-perceived Effects of Lucid Dreaming on Mental and Physical Health," *International Journal of Dream Research*, 2020

Fagiani, Francesca, et al., "The Circadian Molecular Machinery in CNS Cells: A Fine Tuner of Neuronal and Glial Activity with Space/Time Resolution," *Frontiers in Molecular Neuroscience*, July 1, 2022

Fan, Fengmei, et al., "Development of the Default-mode Network during Childhood and Adolescence: A Longitudinal Resting-state fMRI Study," *NeuroImage*, 2021

Fazekas, Peter, Nanay, Bence, and Pearson, Joel, "Offline Perception: An Introduction," *Philosophical Transactions of the Royal Society*, October 28, 2020

Fell, Jürgen, et al., "Human Memory Formation Is Accompanied by Rhinal–Hippocampal Coupling and Decoupling," *Nature Neuroscience*, December 2001

Fennig, S., Salganik, E., and Chayat, M., "Psychotic Episodes and Nightmares: A Case Study," *The Journal of Nervous and Mental Disease*, January 1992

Fenwick, Peter, et al., "Lucid Dreaming: Correspondence between Dreamed and Actual Events in One Subject during REM Sleep," *Biological Psychology*, 1984

Fireman, G. D., Levin, R., and Pope, A. W., "Narrative Qualities of Bad Dreams and Nightmares," *Dreaming*, 2014

Fogel, Stuart M., et al., "A Novel Approach to Dream Content Analysis Reveals Links between Learning-related Dream Incorporation and Cognitive Abilities," *Frontiers in Psychology*, August 8, 2018

Fogli, Alessandro, Aiello, Luca Maria, and Quercia, Daniele, "Our Dreams, Our Selves: Automatic Analysis of Dream Reports," *Royal Society Open Science*, August 26, 2020

Foulkes, David, "Sleep and Dreams. Dream Research: 1953–1993," *Sleep*, 1996

Foulkes, David, et al., "REM Dreaming and Cognitive Skills at Age 5–8: A Cross-sectional Study," *International Journal of Behavioral Development*, 1990

Fox, Kieran C. R., Andrews-Hanna, Jessica R., and Christoff, Kalina, "The Neurobiology of Self-generated Thought from Cells to Systems: Integrating Evidence from Lesion Studies, Human Intracranial Electrophysiology, Neurochemistry, and Neuroendocrinology," *Neuroscience*, 2016

Fox, Kieran C. R., et al., "Changes in Subjective Experience Elicited by Direct Stimulation of the Human Orbitofrontal Cortex," *Neurology*, September 19, 2018

Fox, Kieran C. R., et al., "Dreaming as Mind Wandering: Evidence from Functional Neuroimaging and First-person Content Reports," *Frontiers in Human Neuroscience*, July 30, 2013

Fox, Kieran C. R., et al., "Intrinsic Network Architecture Predicts the Effects Elicited by Intracranial Electrical Stimulation of the Human Brain," *Nature Human Behaviour*, October 2020

Fränkl, Eirin, et al., "How Our Dreams Changed during the COVID-19 Pandemic: Effects and Correlates of Dream Recall Frequency—A Multinational Study on 19,355 Adults," *Nature and Science of Sleep*, 2021

Frick, Andrea, Hansen, Melissa, and Newcombe, Nora S., "Development of Mental Rotation in 3- to 5-year-old Children," *Cognitive Development*, 2013

Fried, Itzhak, et al., "Electric Current Stimulates Laughter," *Nature*, February 12, 1998

Fried, Itzhak, MacDonald, Katherine A., and Wilson, Charles L., "Single Neuron Activity in Human Hippocampus and Amygdala during Recognition of Faces and Objects," *Neuron*, May 1997

Fröhlich, Flavio, Sellers, Kristin K., and Cordle, Asa L, "Targeting the Neurophysiology of Cognitive Systems with Transcranial Alternating Current Stimulation," *Expert Review of Neurotherapeutics*, December 30, 2014

Fulford, Jon, et al., "The Neural Correlates of Visual Imagery Vividness—An fMRI Study and Literature Review," *Cortex*, 2018

Funkhouser, Arthur, "Dreams and Dreaming among the Elderly: An Overview," *Aging and Mental Health*, June 2010

Garcia, Odalis, et al., "What Goes Around Comes Around: Nightmares and Daily Stress Are Bidirectionally Associated in Nurses," *Stress and Health*, 2021

Gauchat, Aline, et al., "The Content of Recurrent Dreams in Young Adolescents," *Consciousness and Cognition*, December 2015

Georgiadis, J. R., and Kringelbach, M. L., "The Human Sexual Response Cycle: Brain Imaging Evidence Linking Sex to Other Pleasures," *Progress in Neurobiology*, 2012

Gerrans, Philip, "Dream Experience and a Revisionist Account of Delusions of Misidentification," *Consciousness and Cognition*, 2012

Gerrans, Philip, "Pathologies of Hyperfamiliarity in Dreams, Delusions, and Déjà Vu," *Frontiers in Psychology*, February 20, 2014

Gieselmann, Annika, et al., "Aetiology and Treatment of Nightmare Disorder: State of the Art and Future Perspectives," *Journal of Sleep Research*, November 22, 2018

Giordano, Alessandra, et al., "Body Schema Self-awareness and Related Dream Content Modifications in Amputees Due to Cancer," *Brain Sciences*, December 9, 2021

Giordano, Alessandra, et al., "Dream Content Changes in Women After Mastectomy: An Initial Study of Body Imagery after Body-disfiguring Surgery," *Dreaming*, 2012

Glasser, Matthew F., et al., "A Multi-modal Parcellation of Human Cerebral Cortex," *Nature*, August 11, 2016

Gofton, Teneille E., et al., "Cerebral Cortical Activity after Withdrawal of Life-sustaining Measures in Critically Ill Patients," *American Journal of Transplantation*, July 13, 2022

Golden, R., et al., "Representation of Memories in an Abstract Synaptic Space and Its Evolution with and without Sleep," *PLOS Computational Biology*, 2022

Golden, Ryan, et al., "Sleep Prevents Catastrophic Forgetting in Spiking Neural Networks by Forming a Joint Synaptic Weight Representation," *PLOS Computational Biology*, 2022

Gomes, Marleide da Mota, and Nardi, Antonio E., "Charles Dickens' Hypnagogia, Dreams, and Creativity," *Frontiers in Psychology*, July 27, 2021

Gorgoni, Maurizio, et al., "Pandemic Dreams: Quantitative and Qualitative Features of the Oneiric Activity during the Lockdown Due to COVID-19 in Italy," *Sleep Medicine*, May 2021

Gott, Jarrod, et al., "Sleep Fragmentation and Lucid Dreaming," *Consciousness and Cognition*, 2020

Gott, Jarrod, et al., "Virtual Reality Training of Lucid Dreaming," *Philosophical Transactions of the Royal Society*, July 13, 2020

Gottesmann, Claude, "The Development of the Science of Dreaming," *International Review of Neurobiology*, 2010

Gottesmann, Claude, "To What Extent Do Neurobiological Sleep-waking Processes Support Psychoanalysis?" *International Review of Neurobiology*, 2010

Goyal, S., et al., "Drugs and Dreams," *Indian Journal of Clinical Practice*, May 2011

Greenberg, Daniel L., and Knowlton, Barbara J., "The Role of Visual Imagery in Autobiographical Memory," *Memory & Cognition*, 2014

Gregor, Thomas, "A Content Analysis of Mehinaku Dreams," *Ethos*, 1981

Griffith, Richard M., Miyagi, Otoya, and Tago, Akira, "The Universality of Typical Dreams: Japanese vs. Americans," *American Anthropologist*, December 1958

Grover, Sandeep, and Mehra, Aseem, "Incubus Syndrome: A Case Series and Review of Literature," *Indian Journal of Psychological Medicine*, 2018

Guillory, Sean A., and Bujarski, Krzysztof A., "Exploring Emotions Using Invasive Methods: Review of 60 Years of Human Intracranial Electrophysiology," *Scan*, 2014

Gulyás, Erzsébet, et al., "Visual Imagery Vividness Declines across the Lifespan," *Cortex*, 2022

Hall, C. S., "Diagnosing Personality by the Analysis of Dreams," *The Journal of Abnormal and Social Psychology*, 1947

Hall, C. S., "What People Dream About," *Scientific American*, May 1951

Hansen, Kathrin, et al., "Efficacy of Psychological Interventions Aiming to Reduce Chronic Nightmares: A Meta-analysis," *Clinical Psychology Review*, February 2013

Harris, Kenneth D., and Thiele, Alexander, "Cortical State and Attention," *Nature Reviews Neuroscience*, September 2011

Hartmann, Ernest, "Making Connections in a Safe Place: Is Dreaming Psychotherapy?" *Dreaming*, 1995

Hartmann, Ernest, "Nightmare after Trauma as Paradigm for All Dreams: A New Approach to the Nature and Functions of Dreaming," *Psychiatry*, 1998

Hartmann, Ernest, "The Underlying Emotion and the Dream: Relating Dream Imagery to the Dreamer's Underlying Emotion Can Help Elucidate the Nature of Dreaming," *International Review of Neurobiology*, 2010

Hartmann, Ernest, et al., "Who Has Nightmares? The Personality of the Lifelong Nightmare Sufferer," *Archives of General Psychiatry*, 1987

Hawkins, G. E., et al., "Toward a Model-based Cognitive Neuroscience of Mind Wandering," *Neuroscience*, 2015

Heather-Greener, Gail Q., Comstock, Dana, and Joyce, Roby, "An Investigation of the Manifest Dream Content Associated with Migraine Headaches: A Study of the Dreams That Precede Nocturnal Migraines," *Psychotherapy and Psychosomatics*, 1996

Hefez, Albert, Metz, Lily, and Lavie, Peretz, "Long-term Effects of Extreme Situational Stress on Sleep and Dreaming," *American Journal of Psychiatry*, 1987

Herlin, Bastien, et al., "Evidence that Non-dreamers Do Dream: A REM Sleep Behaviour Disorder Model," *Journal of Sleep Research*, 2015

Hertenstein, Matthew J., et al., "Touch Communicates Distinct Emotions," *Emotion*, 2006

Hirst, Manton, "Dreams and Medicines: The Perspective of Xhosa Diviners and Novices in the Eastern Cape, South Africa," *Indo-Pacific Journal of Phenomenology*, December 2005

Hobson, Allan, and Kahn, David, "Dream Content: Individual and Generic Aspects," *Consciousness and Cognition*, December 2007

Holzinger, Brigitte, Saletu, Bernd, and Klösch, Gerhard, "Cognitions in Sleep: Lucid Dreaming as an Intervention for Nightmares in Patients with Posttraumatic Stress Disorder," *Frontiers in Psychology*, 2020

Hong, Charles Chong-Hwa, et al., "Rapid Eye Movements in Sleep Furnish a Unique Probe into Consciousness," *Frontiers in Psychology*, October 31, 2018

Hong, Charles Chong-Hwa, Fallon, James H, and Friston, Karl J., "fMRI Evidence for Default Mode Network Deactivation Associated with Rapid Eye Movements in Sleep," *Brain Sciences*, 2021

Horikawa, T., et al., "Neural Decoding of Visual Imagery during Sleep," *Science*, 2013

Hornung, Orla P., "The Relationship between REM Sleep and Memory Consolidation in Old Age and Effects of Cholinergic Medication," *Biological Psychiatry*, 2007

Horton, Caroline L., "Key Concepts in Dream Research: Cognition and Consciousness Are Inherently Linked, but Do No Not Control 'Control'!" *Frontiers in Human Neuroscience*, July 17, 2020

Horváth, Gyöngyvér, "Visual Imagination and the Narrative Image: Parallelisms between Art History and Neuroscience," *Cortex*, 2018

Hoss, Robert J., "Content Analysis on the Potential Significance of Color in Dreams: A Preliminary Investigation," *International Journal of Dream Research*, 2010

Hossain, Shyla R., Simner, Julia, and Ipser, Alberta, "Personality Predicts the Vibrancy of Colour Imagery: The Case of Synaesthesia," *Cortex*, 2018

Inman, Cory S., et al., "Human Amygdala Stimulation Effects on Emotion Physiology and Emotional Experience," *Neuropsychologia*, 2020

Iorio, Ilaria, Sommantico, Massimiliano, and Parrello, Santa, "Dreaming in the Time of COVID-19: A Quali-quantitative Italian Study," *Dreaming*, 2020

Jacobs, Christianne, Schwarzkopf, Dietrich S., and Silvanto, Juha, "Visual Working Memory Performance in Aphantasia," *Cortex*, 2018

Jafari, Eisa, et al., "Intensified Electrical Stimulation Targeting Lateral and Medial Prefrontal Cortices for the Treatment of Social Anxiety Disorder: A Randomized, Double-blind, Parallel-group, Dose-comparison Study," *Brain Stimulation*, 2021

Jalal, Baland, "How to Make the Ghosts in My Bedroom Disappear? Focused-attention Meditation Combined with Muscle Relaxation (MR Therapy)—A Direct Treatment Intervention for Sleep Paralysis," *Frontiers in Psychology*, 2016

Jalal, Baland, "'Men Fear Most What They Cannot See.' Sleep Paralysis 'Ghost Intruders' and Faceless 'Shadow-people'—The Role of the Right Hemisphere and Economizing Nature of Vision," *Medical Hypotheses*, 2021

Jalal, Baland, "The Neuropharmacology of Sleep Paralysis Hallucinations: Serotonin 2A Activation and a Novel Therapeutic Drug," *Psychopharmacology*, 2018

Jalal, Baland, and Hinton, Devon E., "Rates and Characteristics of Sleep Paralysis in the General Population of Denmark and Egypt," *Culture, Medicine and Psychiatry*, 2013

Jalal, Baland, and Ramachandran, Vilayanur S., "Sleep Paralysis and 'the Bedroom Intruder': The Role of the Right Superior Parietal, Phantom Pain and Body Image Projection," *Medical Hypotheses*, 2014

Jalal, Baland, Romanelli, Andrea, and Hinton, Devon E., "Cultural Explanations of Sleep Paralysis in Italy: The Pandafeche Attack and Associated Supernatural Beliefs," *Culture, Medicine and Psychiatry*, March 2015

James, Ella L., et al., "Computer Game Play Reduces Intrusive Memories of Experimental Trauma via Reconsolidation-update Mechanisms," *Psychological Science*, 2015

Janssen, Diederik F., "First Stirrings: Cultural Notes on Orgasm, Ejaculation, and Wet Dreams," *Journal of Sex Research*, 2007

Janszky, J., et al., "Orgasmic Aura—A Report of Seven Cases," *Seizure*, 2004

Jensen, Ole, Kaiser, Jochen, and Lachaux, Jean-Philippe, "Human Gamma-frequency Oscillations Associated with Attention and Memory," *Trends in Neurosciences*, 2007

Jiang, Yi, et al., "A Gender- and Sexual Orientation-dependent Spatial Attentional Effect of Invisible Images," *Proceedings of the National Academy of Sciences*, November 7, 2006

Johnson, E. L., et al., "Direct Brain Recordings Reveal Prefrontal Cortex Dynamics of Memory Development," *Scientific Advances*, 2018

Jun, Jin-Sun, et al., "Emotional and Environmental Factors Aggravating Dream Enactment Behaviors in Patients with Isolated REM Sleep Behavior Disorder," *Nature and Science of Sleep*, September 24, 2022

Jus, A., et al., "Studies on Dream Recall in Chronic Schizophrenic Patients after Frontal Lobotomy," *Biological Psychiatry*, 1973

Kahn David, "Brain Basis of Self: Self-organization and Lessons from Dreaming," *Frontiers in Psychology*, July 16, 2013

Kahn, David, "Reactions to Dream Content: Continuity and Non-continuity," *Frontiers in Psychology*, December 3, 2019

Kahn, David, and Gover, Tzivia, "Consciousness in Dreams," *International Review of Neurobiology*, 2010

Kahn, David, and Hobson, Allan, "Theory of Mind in Dreaming: Awareness of Feelings and Thoughts of Others in Dreams," *Dreaming*, 2005

Kam, Julia W. Y., Mittner, Matthias, and Knight, Robert T., "Mind-wandering: Mechanistic Insights from Lesion, tDCS, and iEEG," *Trends in Cognitive Sciences*, March 2022

Kay, Kenneth, and Frank, Loren, M., "Three Brain States in the Hippocampus and Cortex," *Hippocampus*, 2019

Kellermann, Natan P. F., "Epigenetic Transmission of Holocaust Trauma: Can Nightmares Be Inherited?" *Israel Journal of Psychiatry and Related Sciences*, 2013

Keogh, Rebecca, and Pearson, Joel, "The Blind Mind: No Sensory Visual Imagery in Aphantasia," *Cortex*, 2018

Khambhati, Ankit N., et al., "Functional Control of Electrophysiological Network Architecture Using Direct Neurostimulation in Humans," *Network Neuroscience*, April 14, 2019

King, David B., DeCicco, Teresa L., and Humphreys, Terry P., "Investigating Sexual Dream Imagery in Relation to Daytime Sexual Behaviours and Fantasies among Canadian University Students," *The Canadian Journal of Human Sexuality*, 2009

Kirmayer, Laurence J., "Nightmares, Neurophenomenology and the Cultural Logic of Trauma," *Culture, Medicine and Psychiatry*, 2016

Kleitman, Nathaniel, "Patterns of Dreaming," *Scientific American*, 1960

Komar, Sierra, "Insomniac Technologies: Sleep Wearables Ensure That You Are Never Really at Rest," *Real Life*, April 21, 2022

König, Nina, and Schredl, Michael, "Music in Dreams: A Diary Study," *Psychology of Music*, 2021

Köthe, Martina, and Pietrowsky, Reinhard, "Behavioral Effects of Nightmares and Their Correlations to Personality Patterns," *Dreaming*, 2001

Koutroumanidis, Michael, et al., "Tooth Brushing–induced Seizures: A Case Report," *Epilepsia*, 2001

Krakow, Barry, and Zadra, Antonio, "Clinical Management of Chronic Nightmares: Imagery Rehearsal Therapy," *Behavioral Sleep Medicine*, 2006

Krakow, Barry, et al., "Nightmare Frequency in Sexual Assault Survivors with PTSD," *Journal of Anxiety Disorders*, 2002

Krishnan, Dolly, "Orchestration of Dreams: A Possible Tool for Enhancement of Mental Productivity and Efficiency," *Sleep and Biological Rhythms*, January 2021

Krone, Lukas, et al., "Top-down Control of Arousal and Sleep: Fundamentals and Clinical Implications," *Sleep Medicine Reviews*, 2017

Kroth, Jerry, et al., "Dream Characteristics of Stock Brokers after a Major Market Downturn," *Psychological Reports*, 2002

Kroth, Jerry, et al., "Dream Reports and Marital Satisfaction," *Psychological Reports*, 2005

Kruger, Tyler B., et al., "Using Deliberate Mind-wandering to Escape Negative Mood States: Implications for Gambling to Escape," *Journal of Behavioral Addictions*, October 2, 2020

Ku, Jeonghun, et al., "Brain Mechanisms Involved in Processing Unreal Perceptions," *NeuroImage*, 2008

Kumar, Santosh, Soren, Subhash, and Chaudhury, Suprakash, "Hallucinations: Etiology and Clinical Implications," *Industrial Psychiatry Journal*, 2009

Kunze, Anna E., Arntz, Arnoud, and Kindt, Merel, "Fear Conditioning with Film Clips: A Complex Associative Learning Paradigm," *Journal of Behavior Therapy and Experimental Psychiatry*, 2015

Kunze, Anna E., et al., "Efficacy of Imagery Rescripting and Imaginal Exposure for Nightmares: A Randomized Wait-list Controlled Trial," *Behaviour Research and Therapy*, 2017

Kussé, Caroline, et al., "Neuroimaging of Dreaming: State of the Art and Limitations," *International Review of Neurobiology*, 2010

Kuzmičová, Anežka, "Presence in the Reading of Literary Narrative: A Case for Motor Enactment," *Semiotica*, 2011

LaBerge, Stephen, Baird, Benjamin, and Zimbardo, Philip G., "Smooth Tracking of Visual Targets Distinguishes Lucid REM Sleep Dreaming and Waking Perception from Imagination," *Nature Communications*, 2018

Lai, George, et al., "Acute Effects and the Dreamy State Evoked by Deep Brain Electrical Stimulation of the Amygdala: Associations of the Amygdala in Human Dreaming, Consciousness, Emotions, and Creativity," *Frontiers in Human Neuroscience*, February 25, 2020

Lakoff, George, "How Metaphor Structures Dreams: The Theory of Conceptual Metaphor Applied to Dream Analysis," *Dreaming*, 1993

Lamberg, Lynne, "Scientists Never Dreamed Finding Would Shape a Half-century of Sleep Research," *JAMA*, 2003

Lancee, Jaap, Spoormaker, Victor I., and van den Bout, Jan, "Nightmare Frequency Is Associated with Subjective Sleep Quality but Not with Psychopathology," *Sleep and Biological Rhythms*, 2010

Lancee, Jaap, et al., "A Systematic Review of Cognitive-behavioral Treatment for Nightmares: Toward a Well-established Treatment," *Journal of Clinical Sleep Medicine*, 2008

Landin-Romero, Ramon, et al., "How Does Eye Movement Desensitization and Reprocessing Therapy Work? A Systematic Review on Suggested Mechanisms of Action," *Frontiers in Psychology*, August 13, 2018

Lansky, Melvin R., "Nightmares of a Hospitalized Rape Victim," *Bulletin of the Menninger Clinic*; Winter 1995

Lara-Carrasco, Jessica, et al., "Overnight Emotional Adaptation to Negative Stimuli Is Altered by REM Sleep Deprivation and Is Correlated with Intervening Dream Emotions," *Journal of Sleep Research*, 2009

Lavie, P., et al., "Localized Pontine Lesion: Nearly Total Absence of REM Sleep," *Neurology*, January 1984

Leary, Eileen B., et al., "Association of Rapid Eye Movement Sleep with Mortality in Middle-aged and Older Adults," *JAMA Neurology*, July 6, 2020

Lee, Seung-Hee, and Dan, Yang, "Neuromodulation of Brain States," *Neuron*, October 4, 2012

Lee, UnCheol, et al., "Disruption of Frontal–Parietal Communication by Ketamine, Propofol, and Sevoflurane," *Anesthesiology*, 2013

Leung, Alexander K. C., and Robson, William Lane M., "Nightmares," *Journal of the American Medical Association*, 1993

Levin, Ross, and Nielsen, Tore, "Nightmares, Bad Dreams, and Emotion Dysregulation: A Review and New Neurocognitive Model of Dreaming," *Current Directions in Psychological Science*, 2009

Levin, Ross, and Nielsen, Tore, "Disturbed Dreaming, Posttraumatic Stress Disorder, and Affect Distress: A Review and Neurocognitive Model," *Psychological Bulletin*, 2007

Lewis, J. E., "Dream Reports of Animal Rights Activists," *Dreaming*, 2008

Li, Yanyan, et al., "Neural Substrates of External and Internal Visual Sensations Induced by Human Intracranial Electrical Stimulation," *Frontiers in Neuroscience*, July 2022

Liddon, Sim C., "Sleep Paralysis and Hypnagogic Hallucinations: Their Relationship to the Nightmare," *Archives of General Psychiatry*, 1967

Lima, Susana Q., "Genital Cortex: Development of the Genital Homunculus," *Current Biology*, 2019

Litz, Brett T., et al., "Predictors of Emotional Numbing in Posttraumatic Stress Disorder," *Journal of Traumatic Stress*, 1997

Liu, Siyuan, et al., "Brain Activity and Connectivity during Poetry Composition: Toward a Multidimensional Model of the Creative Process," *Human Brain Mapping*, May 26, 2015

Liu, Xianchen, et al., "Nightmares Are Associated with Future Suicide Attempt and Non-suicidal Self-injury in Adolescents," *Journal of Clinical Psychiatry*, 2019

Livezey, Jeffrey, Oliver, Thomas, and Cantilena, Louis, "Prolonged Neuropsychiatric Symptoms in a Military Service Member Exposed to Mefloquine," *Drug Safety Case Reports*, 2016

Llewellyn, Sue, "Crossing the Invisible Line: De-differentiation of Wake, Sleep and Dreaming May Engender Both Creative Insight and Psychopathology," *Consciousness and Cognition*, 2016

Llewellyn, Sue, "Dream to Predict? REM Dreaming as Prospective Coding," *Frontiers in Psychology*, January 5, 2016

Llewellyn, Sue, and Desseilles, Martin, "Editorial: Do Both Psychopathology and Creativity Result from a Labile Wake–Sleep–Dream Cycle," *Frontiers in Psychology*, October 20, 2017

Lortie-Lussier, Monique, Schwab, Christine, and De Koninck, Joseph, "Working Mothers versus Homemakers: Do Dreams Reflect the Changing Roles of Women?" *Sex Roles*, May 1985

Lusignan, Félix-Antoine, et al., "Dream Content in Chronically-treated Persons with Schizophrenia," *Schizophrenia Research*, 2009

MacKay, Cassidy, and DeCicco, Teresa L., "Pandemic Dreaming: The Effect of COVID-19 on Dream Imagery, a Pilot Study," *Dreaming*, 2020

MacKisack, Matthew, "Painter and Scribe: From Model of Mind to Cognitive Strategy," *Cortex*, 2018

Maggiolini, Alfio, et al., "Typical Dreams across the Life Cycle," *International Journal of Dream Research*, 2020

Magidov, Efrat, et al., "Near-total Absence of REM Sleep Co-occurring with Normal Cognition: An Update of the 1984 Paper," *Sleep Medicine*, 2018

Mahowald, Mark W., and Schenck, Carlos H., "Insights from Studying Human Sleep Disorders," *Nature*, October 27, 2005

Mainieri, Greta, et al., "Are Sleep Paralysis and False Awakenings Different from REM Sleep and from Lucid REM Sleep? A Spectral EEG Analysis," *Journal of Clinical Sleep Medicine*, April 1, 2021

Mallett, Remington, "Partial Memory Reinstatement while (Lucid) Dreaming to Change the Dream Environment," *Consciousness and Cognition*, 2020

Manni, R., and Terzaghi, M., "Dreaming and Enacting Dreams in Non-rapid Eye Movement and Rapid Eye Movement Parasomnia: A Step Toward a Unifying View within Distinct Patterns?" *Sleep Medicine*, 2013

Manni, Raffaele, et al., "Hallucinations and REM Sleep Behaviour Disorder in Parkinson's Disease: Dream Imagery Intrusions and Other Hypotheses," *Consciousness and Cognition*, 2011

Maquet, Pierre, "The Role of Sleep in Learning and Memory," *Science*, 2001

Maquet, Pierre, et al., "Functional Neuroanatomy of Human Rapid-eye-movement Sleep and Dreaming," *Nature*, September 12, 1996

Marinelli, Lydia, "Screening Wish Theories: Dream Psychologies and Early Cinema," *Science in Context*, 2006

Mason, Malia F., et al., "Wandering Minds: The Default Network and Stimulus-independent Thought," *Science*, January 19, 2007

McCaig, R. Graeme, et al., "Improved Modulation of Rostrolateral Prefrontal Cortex Using Real-time fMRI Training and Meta-cognitive Awareness," *NeuroImage*, 2011

McCormick, Cornelia, et al., "Mind-wandering in People with Hippocampal Damage," *The Journal of Neuroscience*, March 14, 2018

McCormick, L., et al., "REM Sleep Dream Mentation in Right Hemispherectomized Patients," *Neuropsychologia*, 1997

McKiernan, Kristen A., et al., "Interrupting the 'Stream of Consciousness': An fMRI Investigation," *NeuroImage*, 2006

McNally, Richard J., and Clancy, Susan A., "Sleep Paralysis, Sexual Abuse, and Space Alien Abduction," *Transcultural Psychiatry*, March 2005

McNamara, Patrick, et al., "Impact of REM Sleep on Distortions of Self-concept, Mood and Memory in Depressed/Anxious Participants," *Journal of Affective Disorders*, 2010

Melzack, Ronald, "Phantom Limbs, the Self and the Brain," *Canadian Psychology*, 1989

Mevel, Katell, et al., "The Default Mode Network in Healthy Aging and Alzheimer's Disease," *International Journal of Alzheimer's Disease*, 2011

Michels, Lars, et al., "The Somatosensory Representation of the Human Clitoris: An fMRI Study," *NeuroImage*, 2010

Mikulincer, Mario, Shaver, Phillip R., and Avihou-Kanza, Neta, "Individual Differences in Adult Attachment Are Systematically Related to Dream Narratives," *Attachment & Human Development*, 2011

Mills, Caitlin, et al., "Is an Off-task Mind a Freely-moving Mind? Examining the Relationship between Different Dimensions of Thought," *Consciousness and Cognition*, 2018

Molendijk, Marc L., et al., "Prevalence Rates of the Incubus Phenomenon: A Systematic Review and Meta-analysis," *Frontiers in Psychiatry*, November 24, 2017

Morewedge, Carey K., and Norton, Michael I., "When Dreaming Is Believing: The (Motivated) Interpretation of Dreams," *Journal of Personality and Social Psychology*, 2009

Mota, Natália B., et al., "Graph Analysis of Dream Reports Is Especially Informative about Psychosis," *Scientific Reports*, January 15, 2014

Mota, Natália B., et al., "Dreaming during the Covid-19 Pandemic: Computational Assessment of Dream Reports Reveals Mental Suffering Related to Fear of Contagion," *PLOS One*, November 30, 2020

Mota-Rolim, Sérgio A., and Araujo, John F., "Neurobiology and Clinical Implications of Lucid Dreaming," *Medical Hypotheses*, 2013

Mota-Rolim, Sérgio A., de Almondes, Katie M., and Kirov, Roumen, "Editorial: 'Is this a Dream?'—Evolutionary, Neurobiological and Psychopathological Perspectives on Lucid Dreaming," *Frontiers in Psychology*, 2021

Mota-Rolim, Sérgio A., et al., "Different Kinds of Subjective Experience during Lucid Dreaming May Have Different Neural Substrates," *International Journal of Dream Research*, 2010

Mota-Rolim, Sérgio A., et al., "Portable Devices to Induce Lucid Dreaming—Are They Reliable?" *Frontiers in Neuroscience*, May 8, 2019

Mota-Rolim, Sérgio A., et al., "The Dream of God: How Do Religion and Science See Lucid Dreaming and Other Conscious States during Sleep?" *Frontiers in Psychology*, October 6, 2020

Moulton, Samuel T., and Kosslyn, Stephen M., "Imagining Predictions: Mental Imagery as Mental Emulation," *Philosophical Transactions of the Royal Society*, 2008

Moyne, Maëva, et al., "Brain Reactivity to Emotion Persists in NREM Sleep and Is Associated with Individual Dream Recall," *Cerebral Cortex Communications*, 2022

Mukamel, Roy, and Fried, Itzhak, "Human Intracranial Recordings and Cognitive Neuroscience," *Annual Review of Psychology*, 2012

Mullally, Sinéad L., and Maguire, Eleanor A., "Memory, Imagination, and Predicting the Future: A Common Brain Mechanism?" *The Neuroscientist*, 2014

Muret, Dollyane, et al., "Beyond Body Maps: Information Content of Specific Body Parts Is Distributed across the Somatosensory Homunculus," *Cell Reports*, 2022

Murzyn, Eva, "Do We Only Dream in Colour? A Comparison of Reported Dream Colour in Younger and Older Adults with Different Experiences of Black and White Media," *Consciousness and Cognition*, 2008

Musse, Fernanda Cristina Coelho, et al., "Mental Violence: The COVID-19 Nightmare," *Frontiers in Psychiatry*, October 30, 2020

Nagy, Tamás, et al., "Frequent Nightmares Are Associated with Blunted Cortisol Awakening Response in Women," *Physiology & Behavior*, 2015

Naiman, Rubin, "Dreamless: The Silent Epidemic of REM Sleep Loss," *Annals of the New York Academy of Sciences*, August 15, 2017

Najam, N., et al., "Dream Content: Reflections of the Emotional and Psychological States of Earthquake Survivors," *Dreaming*, 2006

Nanay, Bence, "Multimodal Mental Imagery," *Cortex*, 2018

Nathan, R. J., Rose-Itkoff, C., and Lord, G., "Dreams, First Memories, and Brain Atrophy in the Elderly," *Hillside Journal of Clinical Psychiatry*, 1981

Neimeyer, Robert A , Torres, Carlos, and Smith, Douglas C., "The Virtual Dream: Rewriting Stories of Loss and Grief," *Death Studies*, 2011

Nemeth, Georgina, "The Route to Recall a Dream: Theoretical Considerations and Methodological Implications," *Psychological Research*, August 12, 2022

Nevin, Remington L., "A Serious Nightmare: Psychiatric and Neurologic Adverse Reactions to Mefloquine Are Serious Adverse Reactions," *Pharmacology Research & Perspectives*, June 5, 2017

Nevin, Remington L., and Ritchie, Elspeth Cameron, "FDA Black Box, VA Red Ink? A Successful Service-connected Disability Claim for Chronic

Neuropsychiatric Adverse Effects from Mefloquine," *Federal Practitioner*, 2016

Nicolas, Alain, and Ruby, Perrine M., "Dreams, Sleep and Psychotropic Drugs," *Frontiers in Neurology*, November 5, 2020

Nielsen, Tore, "Nightmares Associated with the Eveningness Chronotype," *Journal of Biological Rhythms*, February 2010

Nielsen, Tore, "The Stress Acceleration Hypothesis of Nightmares," *Frontiers in Neurology*, June 1, 2017

Nielsen, Tore, and Levin, Ross, "Nightmares: A New Neurocognitive Model," *Sleep Medicine Reviews*, 2007

Nielsen, Tore, and Paquette, Tyna, "Dream-associated Behaviors Affecting Pregnant and Postpartum Women," *Sleep*, 2007

Nielsen, Tore, and Powell, Russell A., "Dreams of the *Rarebit Fiend*: Food and Diet as Instigators of Bizarre and Disturbing Dreams," *Frontiers in Psychology*, February 17, 2015

Nielsen, Tore, et al., "Immediate and Delayed Incorporations of Events into Dreams: Further Replication and Implications for Dream Function," *Journal of Sleep Research*, 2004

Nielsen, Tore, et al., "REM Sleep Characteristics of Nightmare Sufferers before and after REM Sleep Deprivation," *Sleep Medicine*, 2010

Nir, Yuval, and Tononi, Giulio, "Dreaming and the Brain: From Phenomenology to Neurophysiology," *Trends in Cognitive Sciences*, 2010

Nummenmaa, Lauri, et al., "Topography of Human Erogenous Zones," *Archives of Sexual Behavior*, 2016

Nunn, Charles L., and Samson, David R., "Sleep in a Comparative Context: Investigating How Human Sleep Differs from Sleep in Other Primates," *American Journal of Physical Anthropology*, February 14, 2018

O'Callaghan, Claire, Walpola, Ishan C., and Shine, James M., "Neuromodulation of the Mind-wandering Brain State: The Interaction between Neuromodulatory Tone, Sharp Wave-ripples and Spontaneous Thought," *Philosophical Transactions of the Royal Society*, December 14, 2020

Occhionero, Miranda, and Cicogna, Piera Carla, "Autoscopic Phenomena and One's Own Body Representation in Dreams," *Consciousness and Cognition*, 2011

O'Connor, Alison M., and Evans, Angela D., "The Role of Theory of Mind and Social Skills in Predicting Children's Cheating," *Journal of Experimental Child Psychology*, 2019

O'Donnell, Caitlin, et al., "The Role of Mental Imagery in Mood Amplification: An Investigation across Subclinical Features of Bipolar Disorders," *Cortex*, 2018

Olunu, Esther, et al., "Sleep Paralysis, a Medical Condition with a Diverse Cultural Interpretation," *International Journal of Applied and Basic Medical Research*, 2018

Onians, John, "Art, the Visual Imagination and Neuroscience: The Chauvet Cave, Mona Lisa's Smile and Michelangelo's Terribilità," *Cortex*, 2018

Osorio-Forero, Alejandro, et al., "When the Locus Coeruleus Speaks Up in Sleep: Recent Insights, Emerging Perspectives," *International Journal of Molecular Sciences*, 2022

Otaiku, Abidemi I., "Distressing Dreams, Cognitive Decline, and Risk of Dementia: A Prospective Study of Three Population-based Cohorts," *eClinicalMedicine*, September 21, 2022

Otaiku, Abidemi I., "Distressing Dreams and Risk of Parkinson's Disease: A Population-based Cohort Study," *eClinicalMedicine*, June 2022

Otaiku, Abidemi I., "Dream Content Predicts Motor and Cognitive Decline in Parkinson's Disease," *Movement Disorders Clinical Practice*, 2021

Oudiette, Delphine, et al., "Evidence for the Re-enactment of a Recently Learned Behavior during Sleepwalking," *PLOS One*, March 2011

Owczarski, Wojciech, "Dreaming 'the Unspeakable'? How the Auschwitz Concentration Camp Prisoners Experienced and Understood Their Dreams," *Anthropology of Consciousness*, 2020

Pace-Schott, Edward F., "Dreaming as a Storytelling Instinct," *Frontiers in Psychology*, April 2, 2013

Pace-Schott, Edward F., et al., "Effects of Post-exposure Naps on Exposure Therapy for Social Anxiety," *Psychiatry Research*, October 9, 2018

Pagel, James F., "Post-Freudian PTSD: Breath, the Protector of Dreams," *Journal of Clinical Sleep Medicine*, October 15, 2017

Pagel, James F., "What Physicians Need to Know about Dreams and Dreaming," *Current Opinion in Pulmonary Medicine*, 2012

Pagel, J. F., Kwiatkowski, C., and Broyles, K. E., "Dream Use in Film Making," *Dreaming*, 1999

Paiva, Teresa, Bugalho, Paulo, and Bentes, Carla, "Dreaming and Cognition in Patients with Frontotemporal Dysfunction," *Consciousness and Cognition*, 2011

Palermo, Liana, et al., "Congenital Lack and Extraordinary Ability in Object and Spatial Imagery: An Investigation on Sub-types of Aphantasia and Hyperphantasia," *Consciousness and Cognition*, 2022

Paller, Ken A., Creery, Jessica D., and Schechtman, Eitan, "Memory and Sleep: How Sleep Cognition Can Change the Waking Mind for the Better," *Annual Review of Psychology*, 2021

Parvizi, Josef, "Corticocentric Myopia: Old Bias in New Cognitive Sciences," *Trends in Cognitive Sciences*, 2009

Pearson, Joel, "The Human Imagination: The Cognitive Neuroscience of Visual Mental Imagery," *Nature Reviews Neuroscience*, October 2019

Pearson, Joel, and Westbrook, Fred, "Phantom Perception: Voluntary and Involuntary Nonretinal Vision," *Trends in Cognitive Sciences*, May 2015

Peng, Ke, et al., "Brodmann Area 10: Collating, Integrating and High Level Processing of Nociception and Pain," *Progress in Neurobiology*, December 2017

Perogamvros, L., et al., "Sleep and Dreaming Are for Important Matters," *Frontiers in Psychology*, July 25, 2013

Pesonen, Anu-Katriina, et al., "Pandemic Dreams: Network Analysis of Dream Content during the COVID-19 Lockdown," *Frontiers in Psychology*, October 1, 2020

Picard-Deland, Claudia, et al., "Flying Dreams Stimulated by an Immersive Virtual Reality Task," *Consciousness and Cognition*, 2020

Picard-Deland, Claudia, et al., "The Memory Sources of Dreams: Serial Awakenings across Sleep Stages and Time of Night," *Sleep*, December 3, 2022

Picard-Deland, Claudia, et al., "Whole-body Procedural Learning Benefits from Targeted Memory Reactivation in REM Sleep and Task-related Dreaming," *Neurobiology of Learning and Memory*, 2021

Picchioni, Dante, et al., "Nightmares as a Coping Mechanism for Stress," *Dreaming*, 2002

Plazzi, Giuseppe, "Dante's Description of Narcolepsy," *Sleep Medicine*, 2013

Postuma, Ronald B., et al., "Antidepressants and REM Sleep Behavior Disorder: Isolated Side Effect or Neurodegenerative Signal?" *Sleep*, 2013

Prince, Luke Y., and Richards, Blake A., "The Overfitted Brain Hypothesis," *Patterns*, May 14, 2021

Puig, M. Victoria, and Gulledge, Allan, "Serotonin and Prefrontal Cortex Function: Neurons, Networks, and Circuits," *Molecular Neurobiology*, 2011

Pyasik, Maria, et al., "Shared Neurocognitive Mechanisms of Attenuating Self-touch and Illusory Self-touch," *Social Cognitive and Affective Neuroscience*, 2019

Radziun, Dominika, and Ehrsson, H. Henrik, "Short-term Visual Deprivation Boosts the Flexibility of Body Representation," *Scientific Reports*, April 19, 2018

Raichle, Marcus E., et al., "A Default Mode of Brain Function," *Proceedings of the National Academy of Sciences*, January 16, 2001

Ramachandran, V. S., Rogers-Ramachandran, D., and Stewart, M., "Perceptual Correlates of Massive Cortical Reorganization," *Science*, November 13, 1992

Ramezani, Mahtab, et al., "The Impact of Brain Lesions on Sexual Dysfunction in Patients with Multiple Sclerosis: A Systematic Review of Magnetic Resonance Imaging Studies," *Multiple Sclerosis and Related Disorders*, October 31, 2021

Reid, Sandra D., and Simeon, Donald T., "Progression of Dreams of Crack Cocaine Abusers as a Predictor of Treatment Outcome: A Preliminary Report," *The Journal of Nervous and Mental Disease*, December 2001

Resnick, Jody, et al., "Self-representation and Bizarreness in Children's Dream Reports Collected in the Home Setting," *Consciousness and Cognition*, March 1994

Revonsuo, Antti, "The Reinterpretation of Dreams: An Evolutionary Hypothesis of the Function of Dreaming," *Behavioral and Brain Sciences*, 2000

Rigon, Arianna, et al., "Traumatic Brain Injury and Creative Divergent Thinking," *Brain Injury*, April 2020

Rimsh, A., and Pietrowsky, R., "Analysis of Dream Contents of Patients with Anxiety Disorders and Their Comparison with Dreams of Healthy Participants," *Dreaming*, 2021

Riva, Michele Augusto, et al., "The Neurologist in Dante's *Inferno*," *European Neurology*, April 22, 2015

Rizzolatti, Giacomo, and Arbib, Michael, "Language within Our Grasp," *Trends in Neuroscience*, 1998

Rizzolatti, Giacomo, Fogassi, Leonardo, and Gallese, Vittorio, "Neurophysiological Mechanisms Underlying the Understanding and Imitation of Action," *Nature Reviews Neuroscience*, September 2001

Rosen, Melanie G., "How Bizarre? A Pluralist Approach to Dream Content," *Consciousness and Cognition*, 2018

Ruby, Perrine, et al., "Dynamics of Hippocampus and Orbitofrontal Cortex Activity during Arousing Reactions from Sleep: An Intracranial Electroencephalographic Study," *Human Brain Mapping*, 2021

Russell, Kirsten, et al., "Sleep Problem, Suicide and Self-harm in University Students: A Systematic Review," *Sleep Medicine Reviews*, 2019

Sadavoy, Joel, "Survivors: A Review of the Late-life Effects of Prior Psychological Trauma," *The American Journal of Geriatric Psychiatry*, 1997

Sagnier, S., et al., "Lucid Dreams, an Atypical Sleep Disturbance in Anterior and Mediodorsal Thalamic Strokes," *Revue Neurologique*, 2015

Sanders, K. E. G., et al., "Corrigendum: Targeted Memory Reactivation during Sleep Improves Next-day Problem Solving," *Psychological Science*, 2020

Sándor, Piroska, Szakadát, Sára, and Bódizs, Róbert, "Ontogeny of Dreaming: A Review of Empirical Studies," *Sleep Medicine Reviews*, 2014

Sato, João Ricardo, et al., "Age Effects on the Default Mode and Control Networks in Typically Developing Children," *Journal of Psychiatric Research*, July 18, 2014

Saunders, David T., et al., "Lucid Dreaming Incidence: A Quality Effects Meta-analysis of 50 Years of Research," *Consciousness and Cognition*, 2016

Sbarra, David A., Hasselmo, Karen, and Bourassa, Kyle J., "Divorce and Health: Beyond Individual Differences," *Current Directions in Psychological Science*, 2015

Scarpelli, Serena, et al., "Dreams and Nightmares during the First and Second Wave of the COVID-19 Infection: A Longitudinal Study," *Brain Sciences*, October 20, 2021

Scarpelli, Serena, et al., "Investigation on Neurobiological Mechanisms of Dreaming in the New Decade," *Brain Sciences*, February 11, 2021

Scarpelli, Serena, et al., "Nightmares in People with COVID-19: Did Coronavirus Infect Our Dreams?" *Nature and Science of Sleep*, January 24, 2022

Scarpelli, Serena, et al., "Predicting Dream Recall: EEG Activation during NREM Sleep or Shared Mechanisms with Wakefulness?" *Brain Topography*, April 22, 2017

Scarpelli, Serena, et al., "The Impact of the End of COVID Confinement on Pandemic Dreams, as Assessed by a Weekly Sleep Diary: A Longitudinal Investigation in Italy," *Journal of Sleep Research*, July 20, 2021

Schädlich, Melanie, and Erlacher, Daniel, "Practicing Sports in Lucid Dreams—Characteristics, Effects, and Practical Implications," *Current Issues in Sport Science*, 2018

Schierenbeck, Thomas, et al., "Effect of Illicit Recreational Drugs Upon Sleep: Cocaine, Ecstasy and Marijuana," *Sleep Medicine Reviews*, 2008

Schott, G. D., "Penfield's Homunculus: A Note on Cerebral Cartography," *Journal of Neurology, Neurosurgery and Psychiatry*, April 1993

Schredl, Michael, "Characteristics and Contents of Dreams," *International Review of Neurobiology*, 2010

Schredl, Michael, "Dreams in Patients with Sleep Disorders," *Sleep Medicine Reviews*, 2009

Schredl, Michael, "Explaining the Gender Difference in Nightmare Frequency," *The American Journal of Psychology*, 2014

Schredl, Michael, "Nightmares as a Paradigm for Studying the Effects of Stressors," *Sleep*, July 2013

Schredl, Michael, "Nightmare Frequency and Nightmare Topics in a Representative German Sample," *European Archives of Psychiatry and Clinical Neuroscience*, 2010

Schredl, Michael, "Reminiscences of Love: Former Romantic Partners in Dreams," *International Journal of Dream Research*, 2018

Schredl, Michael, and Bulkeley, Kelly, "Dreaming and the COVID-19 Pandemic: A Survey in a U.S. Sample," *Dreaming*, 2020

Schredl, Michael, and Erlacher, Daniel, "Fever Dreams: An Online Study," *Frontiers in Psychology*, January 28, 2020

Schredl, Michael, and Erlacher, Daniel, "Relation between Waking Sport Activities, Reading, and Dream Content in Sport Students and Psychology Students," *The Journal of Psychology*, 2008

Schredl, Michael, and Göritz, Anja S., "Nightmares, Chronotype, Urbanicity, and Personality: An Online Study," *Clocks & Sleep*, 2020

Schredl, Michael, and Göritz, Anja S., "Nightmare Themes: An Online Study of Most Recent Nightmares and Childhood Nightmares," *Journal of Clinical Sleep Medicine*, March 15, 2018

Schredl, Michael, and Mathes, Jonas, "Are Dreams of Killing Someone Related to Waking-life Aggression?" *Dreaming*, September 2014

Schredl, Michael, and Reinhard, Iris, "Gender Differences in Nightmare Frequency: A Meta-analysis," *Sleep Medicine Reviews*, 2011

Schredl, Michael, and Wood, Lara, "Partners and Ex-partners in Dreams: A Diary Study," *Clocks & Sleep*, May 26, 2021

Schredl, Michael, et al., "Dream Recall, Nightmare Frequency, and Nocturnal Panic Attacks in Patients with Panic Disorder," *The Journal of Nervous and Mental Disease*, August 2001

Schredl, Michael, et al., "Dreaming about Cats: An Online Survey," *Dreaming*, September 13, 2021

Schredl, Michael, et al., "Erotic Dreams and Their Relationship to Waking-life Sexuality," *Sexologies*, June 24, 2008

Schredl, Michael, et al., "Information Processing during Sleep: The Effect of Olfactory Stimuli on Dream Content and Dream Emotions," *Journal of Sleep Research*, 2009

Schredl, Michael, et al., "Nightmare Frequency in Last Trimester of Pregnancy," *BMC Pregnancy and Childbirth*, 2016

Schredl, Michael et al., "Work-related Dreams: An Online Survey," *Clocks & Sleep*, 2020

Schredl, Michael, Funhouser, Arthur, and Arn, Nichole, "Dreams of Truck Drivers: A Test of the Continuity Hypothesis of Dreaming," *Imagination, Cognition and Personality*, 2005

Schwartz, Sophie, Clerget, Alice, and Perogamvros, Lampros, "Enhancing Imagery Rehearsal Therapy for Nightmares with Targeted Memory Reactivation," *Current Biology*, 2022

Selimbeyoglu, Aslihan, and Parvizi, Josef, "Electrical Stimulation of the Human Brain: Perceptual and Behavioral Phenomena Reported in the Old and New Literature," *Frontiers in Human Neuroscience*, May 31, 2010

Selterman, Dylan, Apetroaia, Adela, and Waters, Everett, "Script-like Attachment Representations in Dreams Containing Current Romantic Partners," *Attachment & Human Development*, 2012

Selterman, Dylan, et al., "Dreaming of You: Behavior and Emotion in Dreams of Significant Others Predict Subsequent Relational Behavior," *Social Psychological and Personality Science*, 2014

Serpe, Alexis, and DeCicco, Teresa L., "An Investigation into Anxiety and Depression in Dream Imagery: The Issue of Co-morbidity," *International Journal of Dream Research*, 2020

Serper, Zvika, "Kurosawa's 'Dreams': A Cinematic Reflection of a Traditional Japanese Context," *Cinema Journal*, 2001

Sharpless, Brian A., and Doghramji, Karl, "Commentary: How to Make the Ghosts in My Bedroom Disappear? Focused-attention Meditation Combined with Muscle Relaxation (MR Therapy)—A Direct Treatment Intervention for Sleep Paralysis," *Frontiers in Psychology*, April 3, 2017

Shen, Ying, et al., "Emergence of Sexual Dreams and Emission Following Deep Transcranial Magnetic Stimulation over the Medial Prefrontal and Cingulate Cortices," *CNS & Neurological Disorders—Drug Targets*, 2021

Siclari, Francesca, et al., "The Neural Correlates of Dreaming," *Nature Neuroscience*, April 10, 2017

Siegel, J. M., "The REM Sleep-memory Consolidation Hypothesis," *Science*, 2001

Sikka, Pilleriin, et al., "EEG Frontal Alpha Asymmetry and Dream Affect: Alpha Oscillations Over the Right Frontal Cortex during REM Sleep and Presleep Wakefulness Predict Anger in REM Sleep Dreams," *The Journal of Neuroscience*, June 12, 2019

Simard, Valérie, et al., "Longitudinal Study of Bad Dreams in Preschool-aged Children: Prevalence, Demographic Correlates, Risk and Protective Factors," *Sleep*, 2008

Simor, Péter, et al., "Electroencephalographic and Autonomic Alterations in Subjects with Frequent Nightmares during Pre- and Post-REM Periods," *Brain and Cognition*, 2014

Simor, Péter, et al., "Impaired Executive Functions in Subjects with Frequent Nightmares as Reflected by Performance in Different Neuropsychological Tasks," *Brain and Cognition*, 2012

Singh, Arun, et al., "Evoked Midfrontal Activity Predicts Cognitive Dysfunction in Parkinson's Disease," *MedRxIV*, 2022

Singh, Shantanu, et al., "Parasomnias: A Comprehensive Review," *Cureus*, December 31, 2018

Smallwood, Jonathan, and Schooler, Jonathan W., "The Science of Mind Wandering: Empirically Navigating the Stream of Consciousness," *Annual Review of Psychology*, 2015

Smith, Carlyle, and Newfield, Donna-Marie, "Content Analysis of the Dreams of a Medical Intuitive," *Explore*, 2022

Smith, R. C., "A Possible Biologic Role of Dreaming," *Psychotherapy and Psychosomatics*, 1984

Smith, R. C., "Do Dreams Reflect a Biological State?" *The Journal of Nervous and Mental Disease*, 1987

Solms, Mark, "Dreaming and REM Sleep Are Controlled by Different Brain Mechanisms," *Behavioral and Brain Sciences*, 2000

Solomonova, Elizaveta, et al., "Stuck in a Lockdown: Dreams, Bad Dreams, Nightmares, and Their Relationship to Stress, Depression and Anxiety during the COVID-19 Pandemic," *PLOS One*, November 24, 2021

Song, Tian-He, et al., "Nightmare Distress as a Risk Factor for Suicide among Adolescents with Major Depressive Disorder," *Nature and Science of Sleep*, September 2022

Spanò, Goffredina, et al., "Dreaming with Hippocampal Damage," *eLife*, 2020

Sparrow, Gregory, et al., "Exploring the Effects of Galantamine Paired with Meditation and Dream Reliving on Recalled Dreams: Toward an Integrated Protocol for Lucid Dream Induction and Nightmare Resolution," *Consciousness and Cognition*, 2018

Speth, Jana, Frenzel, Clemens, and Voss, Ursula, "A Differentiating Empirical Linguistic Analysis of Dreamer Activity in Reports of EEG-controlled REM-dreams and Hypnagogic Hallucinations," *Consciousness and Cognition*, 2013

Spoormaker, Victor I., "A Cognitive Model of Recurrent Nightmares," *International Journal of Dream Research*, 2008

Spoormaker, Victor I., and van den Bout, Jan, "Lucid Dreaming Treatment for Nightmares: A Pilot Study," *Psychotherapy and Psychosomatics*, 2006

Spoormaker, Victor I., Schredl, Michael, and van den Bout, Jan, "Nightmares: From Anxiety Symptom to Sleep Disorder," *Sleep Medicine Reviews*, 2006

Spoormaker, Victor I., van den Bout, Jan, and Meijer, Eli J. G., "Lucid Dreaming Treatment for Nightmares: A Series of Cases," *Dreaming*, 2003

Sridharan, Devarajan, Levitin, Daniel J., and Menon, Vinod, "A Critical Role for the Right Fronto-insular Cortex in Switching between Central-executive and Default-mode Networks," *Proceedings of the National Academy of Sciences*, August 26, 2008

Stallman, Helen M., Kohler, Mark, and White, Jason, "Medication Induced Sleepwalking: A Systematic Review," *Sleep Medicine Reviews*, 2018

Staunton, Hugh, "The Function of Dreaming," *Reviews in the Neurosciences*, 2001

Sterpenich, Virginie, et al., "Fear in Dreams and in Wakefulness: Evidence for Day/Night Affective Homeostasis," *Human Brain Mapping*, 2020

Stickgold, Robert, Zadra, Antonio, and Haar, AJH, "Advertising in Dreams Is Coming: Now What?" *DxE*, June 8, 2021

Stocks, Abigail, et al., "Dream Lucidity Is Associated with Positive Waking Mood," *Consciousness and Cognition*, 2020

Stuck, B. A., et al., "Chemosensory Stimulation during Sleep—Arousal Responses to Gustatory Stimulation," *Neuroscience*, 2016

Stumbrys, Tadas, "The Luminous Night of the Soul: The Relationship between Lucid Dreaming and Spirituality," *International Journal of Transpersonal Studies*, 2021

Stumbrys, Tadas, and Daniels, Michael, "An Exploratory Study of Creative Problem Solving in Lucid Dreams: Preliminary Findings and Methodological Considerations," *International Journal of Dream Research*, 2010

Stumbrys, Tadas, and Erlacher, Daniel, "Applications of Lucid Dreams and Their Effects on the Mood Upon Awakening," *International Journal of Dream Research*, 2016

Stumbrys, Tadas, Erlacher, Daniel, and Schredl, Michael, "Effectiveness of Motor Practice in Lucid Dreams: A Comparison with Physical and Mental Practice," *Journal of Sports Sciences*, 2016

Stumbrys, Tadas, Erlacher, Daniel, and Schredl, Michael, "Testing the Involvement of the Prefrontal Cortex in Lucid Dreaming: A tDCS Study," *Consciousness and Cognition*, 2013

Stumbrys, Tadas, et al., "Induction of Lucid Dreams: A Systematic Review of Evidence," *Consciousness and Cognition*, 2012

Stumbrys, Tadas, et al., "The Phenomenology of Lucid Dreaming: An Online Survey," *The American Journal of Psychology*, Summer 2014

Suarez, Ralph O., et al., "Contributions to Singing Ability by the Posterior Portion of the Superior Temporal Gyrus of the Non-language-dominant Hemisphere: First Evidence from Subdural Cortical Stimulation, Wada Testing, and fMRI," *Cortex*, 2010

Szabadi, Elemer, Reading, Paul James, and Pandi-Perumal, Seithikurippu R., "Editorial: The Neuropsychiatry of Dreaming: Brain Mechanisms and Clinical Presentations," *Frontiers in Neurology*, March 25, 2021

Szczepanski, Sara, and Knight, Robert, "Insights into Human Behavior from Lesions to the Prefrontal Cortex," *Neuron*, September 3, 2014

Tallon, Kathleen, et al., "Mental Imagery in Generalized Anxiety Disorder: A Comparison with Healthy Control Participants," *Behaviour Research and Therapy*, 2020

Tan, Shuyue, and Fan, Jialin, "A Systematic Review of New Empirical Data on Lucid Dream Induction Techniques," *Journal of Sleep Research*, November 21, 2022

Titus, Caitlin E., et al., "What Role Do Nightmares Play in Suicide? A Brief Exploration," *Current Opinion in Psychology*, 2018

Torontali, Zoltan A., et al., "The Sublaterodorsal Tegmental Nucleus Functions to Couple Brain State and Motor Activity during REM Sleep and Wakefulness," *Current Biology*, November 18, 2019

Tribl, Gotthard G., et al., "Dream Reflecting Cultural Contexts: Comparing Brazilian and German Diary Dreams and Most Recent Dreams," *International Journal of Dream Research*, 2018

Tribl, Gotthard G., Wetter, Thomas C., and Schredl, Michael, "Dreaming Under Antidepressants: A Systematic Review on Evidence in Depressive Patients and Healthy Volunteers," *Sleep Medicine Reviews*, 2013

Trottia, Lynn Marie, et al., "Cerebrospinal Fluid Hypocretin and Nightmares in Dementia Syndromes," *Dementia and Geriatric Cognitive Disorders Extra*, 2021

Tselebis, Athanasios, Zoumakis, Emmanouil, and Ilias, Ioannis, "Dream Recall/Affect and the Hypothalamic–Pituitary–Adrenal Axis," *Clocks & Sleep*, July 22, 2021

Uguccioni, Ginevra, et al., "Fight or Flight? Dream Content during Sleepwalking/Sleep Terrors vs Rapid Eye Movement Sleep Behavior Disorder," *Sleep Medicine*, 2013

Uitermarkt, Brandt, et al., "Rapid Eye Movement Sleep Patterns of Brain Activation and Deactivation Occur within Unique Functional Networks," *Human Brain Mapping*, June 23, 2020

Ünal, Gülten, and Hohenberger, Annette, "The Cognitive Bases of the Development of Past and Future Episodic Cognition in Preschoolers," *Journal of Experimental Child Psychology*, June 20, 2017

Vaca, Guadalupe Fernández-Baca, et al., "Mirth and Laughter Elicited during Brain Stimulation," *Epileptic Disorders*, 2011

Vaillancourt-Morel, Marie-Pier, et al., "Targets of Erotic Dreams and Their Associations with Waking Couple and Sexual Life," *Dreaming*, 2021

Vallat, Raphael, et al., "High Dream Recall Frequency Is Associated with Increased Creativity and Default Mode Network Connectivity," *Nature and Science of Sleep*, February 22, 2022

Valli, Katja and Revonsuo, Antti, "The Threat Simulation Theory in Light of Recent Empirical Evidence: A Review," *The American Journal of Psychology*, 2009

Valli, Katja, et al., "Dreaming Furiously? A Sleep Laboratory Study on the Dream Content of People with Parkinson's Disease and with or without Rapid Eye Movement Sleep Behavior Disorder," *Sleep Medicine*, 2015

Valli, Katja, et al., "The Threat Simulation Theory of the Evolutionary Function of Dreaming: Evidence from Dreams of Traumatized Children," *Consciousness and Cognition*, 2005

van Gaal, Simon, et al., "Unconscious Activation of the Prefrontal No-go Network," *The Journal of Neuroscience*, March 17, 2010

van Liempt, Saskia, et al., "Impact of Impaired Sleep on the Development of PTSD Symptoms in Combat Veterans: A Prospective Longitudinal Cohort Study," *Depression and Anxiety*, 2013

van Rijn, Elaine, et al., "Daydreams Incorporate Recent Waking Life Concerns but Do Not Show Delayed ('Dream-lag') Incorporations," *Consciousness and Cognition*, 2018

van Rijn, Elaine, et al., "The Dream-lag Effect: Selective Processing of Personally Significant Events during Rapid Eye Movement Sleep, but Not during Slow Wave Sleep," *Neurobiology of Learning and Memory*, 2015

Versace, Francesco, et al., "Brain Responses to Erotic and Other Emotional Stimuli in Breast Cancer Survivors with and without Distress about Low Sexual Desire: A Preliminary fMRI Study," *Brain Imaging and Behavior*, December 2013

Vetrugno, Roberto, Arnulf, Isabelle, and Montagna, Pasquale, "Disappearance of 'Phantom Limb' and Amputated Arm Usage during Dreaming in REM Sleep Behaviour Disorder," *British Medical Journal Case Reports*, 2009

Vicente, Raul, et al., "Enhanced Interplay of Neuronal Coherence and Coupling in the Dying Human Brain," *Frontiers in Aging Neuroscience*, February 22, 2022

Vignal, Jean-Pierre, et al., "The Dreamy State: Hallucinations of Autobiographic Memory Evoked by Temporal Lobe Stimulations and Seizures," *Brain*, 2007

Vitali, Helene, et al., "The Vision of Dreams: From Ontogeny to Dream Engineering in Blindness," *Journal of Clinical Sleep Medicine*, August 1, 2022

Voss, Ursula, et al., "Induction of Self Awareness in Dreams Through Frontal Low Current Stimulation of Gamma Activity," *Nature Neuroscience*, 2014

Voss, Ursula, et al., "Lucid Dreaming: A State of Consciousness with Features of Both Waking and Non-lucid Dreaming," *Sleep*, 2009

Voss, Ursula, et al., "Waking and Dreaming: Related but Structurally Independent. Dream Reports of Congenitally Paraplegic and Deaf–Mute Persons," *Consciousness and Cognition*, 2011

Walker, Matthew P., "Sleep-dependent Memory Processing," *Harvard Review of Psychiatry*, 2008

Wamsley, Erin, "Dreaming and Offline Memory Consolidation," *Current Neurology and Neuroscience Reports*, 2014

Wamsley, Erin, et al., "Delusional Confusion of Dreaming and Reality in Narcolepsy," *Sleep*, February 2014

Wang, Jia Xi, et al., "A Paradigm for Matching Waking Events into Dream Reports," *Frontiers in Psychology*, July 3, 2020

Wang, Jia Xi, and Shen, He Yong, "An Attempt at Matching Waking Events into Dream Reports by Independent Judges," *Frontiers in Psychology*, April 6, 2018

Ward, Amanda M., "A Critical Evaluation of the Validity of Episodic Future Thinking: A Clinical Neuropsychology Perspective," *Neuropsychology*, 2016

Wassing, Rick, et al., "Restless REM Sleep Impedes Overnight Amygdala Adaptation," *Current Biology*, 2019

Watanabe, Takamitsu, "Causal Roles of Prefrontal Cortex during Spontaneous Perceptual Switching Are Determined by Brain State Dynamics," *eLife*, 2021

Waters, Flavie, Barnby, Joseph M., and Blom, Jan Dirk, "Hallucination, Imagery, Dreaming: Reassembling Stimulus-independent Perceptions Based on Edmund Parish's Classic Misperception Framework," *Philosophical Transactions of the Royal Society*, 2020

Waters, Flavie, et al., "What Is the Link between Hallucinations, Dreams, and Hypnagogic–Hypnopompic Experiences?" *Schizophrenia Bulletin*, 2016

Watkins, Nicholas W., "(A)phantasia and Severely Deficient Autobiographical Memory: Scientific and Personal Perspectives," *Cortex*, 2018

Wicken, Marcus, Keogh, Rebecca, and Pearson, Joel, "The Critical Role of Mental Imagery in Human Emotion: Insights from Fear-based Imagery and Aphantasia," *Proceedings of the Royal Society B*, 2021

Windt, Jennifer M., and Noreika, Valdas, "How to Integrate Dreaming into a General Theory of Consciousness—A Critical Review of Existing Positions and Suggestions for Future Research," *Consciousness and Cognition*, 2011

Winlove, Crawford I. P., et al., "The Neural Correlates of Visual Imagery: A Co-ordinate-based Meta-analysis," *Cortex*, 2018

Wittmann, Lutz, Schredl, Michael, and Kramer, Milton, "Dreaming in Posttraumatic Stress Disorder: A Critical Review of Phenomenology, Psychophysiology and Treatment," *Psychotherapy and Psychosomatics*, 2007

Wright, Scott T., et al., "The Impact of Dreams of the Deceased on Bereavement: A Survey of Hospice Caregivers," *American Journal of Hospice and Palliative Medicine*, 2014

Wyatt, Richard J., et al., "Total Prolonged Drug-induced REM Sleep Suppression in Anxious-depressed Patients," *Archives of General Psychiatry*, 1971

Yamaoka, Akina, and Yukawa, Shintaro, "Does Mind Wandering during the Thought Incubation Period Improve Creativity and Worsen Mood?" *Psychological Reports*, October 2020

Yamazaki, Risa, et al., "Evolutionary Origin of NREM and REM Sleep," *Frontiers in Psychology*, 2020

Yin, F., et al., "Typical Dreams of 'Being Chased': A Cross-cultural Comparison between Tibetan and Han Chinese Dreamers," *Dreaming*, 2013

Yu, Calvin Kai-Ching, "Can Students' Dream Experiences Reflect Their Performance in Public Examinations?" *International Journal of Dream Research*, 2016

Yu, Calvin Kai-Ching, "Imperial Dreams and Oneiromancy in Ancient China—We Share Similar Dream Motifs with Our Ancestors Living Two Millennia Ago," *Dreaming*, 2022

Yu, Calvin Kai-Ching, and Fu, Wai, "Sex Dreams, Wet Dreams, and Nocturnal Emissions," *Dreaming*, 2011

Zadra, Antonio, Pilon, Mathieu, and Donderi, Don C., "Variety and Intensity of Emotions in Nightmares and Bad Dreams," *The Journal of Nervous and Mental Disease*, April 2006

Zeman, Adam, et al., "Phantasia—The Psychological Significance of Lifelong Visual Imagery Vividness Extremes," *Cortex*, 2020

Zeman, Adam, MacKisack, Matthew, and Onians, John, "The Eye's Mind—Visual Imagination, Neuroscience and the Humanities," *Cortex*, 2018

Zink, Nicolas, and Pietrowsky, Reinhard, "Relationship between Lucid Dreaming, Creativity and Dream Characteristics," *International Journal of Dream Research*, 2013

Index